BURNING ISSUES

BURNING ISSUES

Fire in Art and the Social Imagination

Alan Krell

REAKTION BOOKS

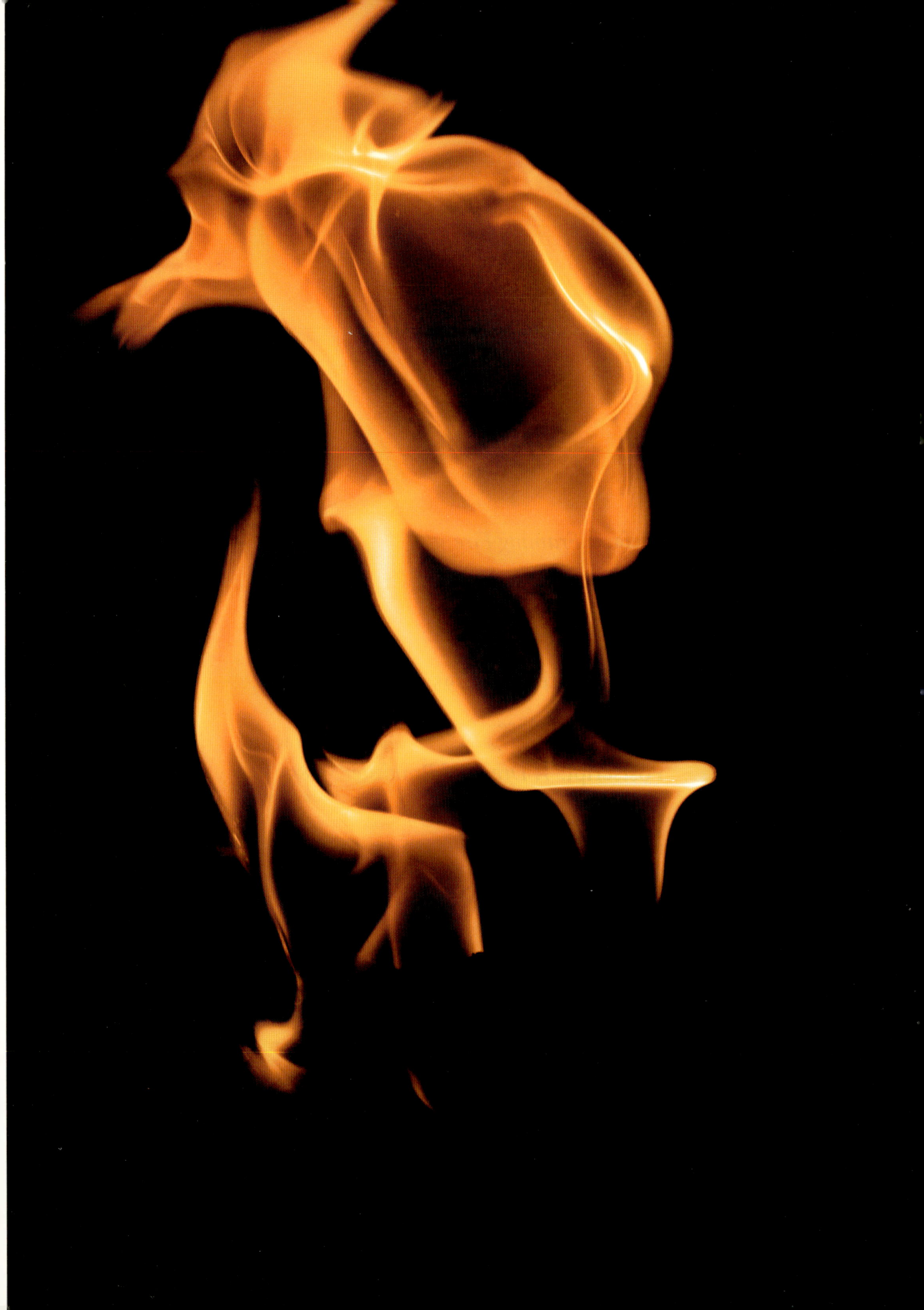

Preface

Framing Fire

Fire's history is contradictory and paradoxical. There's the unassuming fire that results from burning kindling and invites reverie, or the conflagration brought about by the gesture of an arsonist. Fire may be the inadvertent spark that results in destruction or the wholesale damage brought about by military might. Yet we cook with fire and warm ourselves in front of its comforting presence. Ravaging the environment, it also replenishes. Fire destroys lives and structures, but it also mesmerizes and we celebrate it. Damning yet redemptive, destructive but also comforting, fire leads a charmed life.

This Janus-like character of fire is the focus of *Burning Issues*. By examining engagements with fire that variously document, re-locate and mythologize, the book covers commentary in journals and newspapers, religious stories and rituals, diaries, paintings, photographs and children's tales; it concludes with installation and performance works from the mid-twentieth century onwards that make use of actual fire. Questions about the private and the public, the intimate and the invasive, the destructive and the redemptive, weave in and out of the text.

I am in the J. Paul Getty Museum high up in the hills surrounding Los Angeles, looking at a very small image of fire painted by the Flemish artist William Vrelant or his workshop in the early 1460s: it's on a page

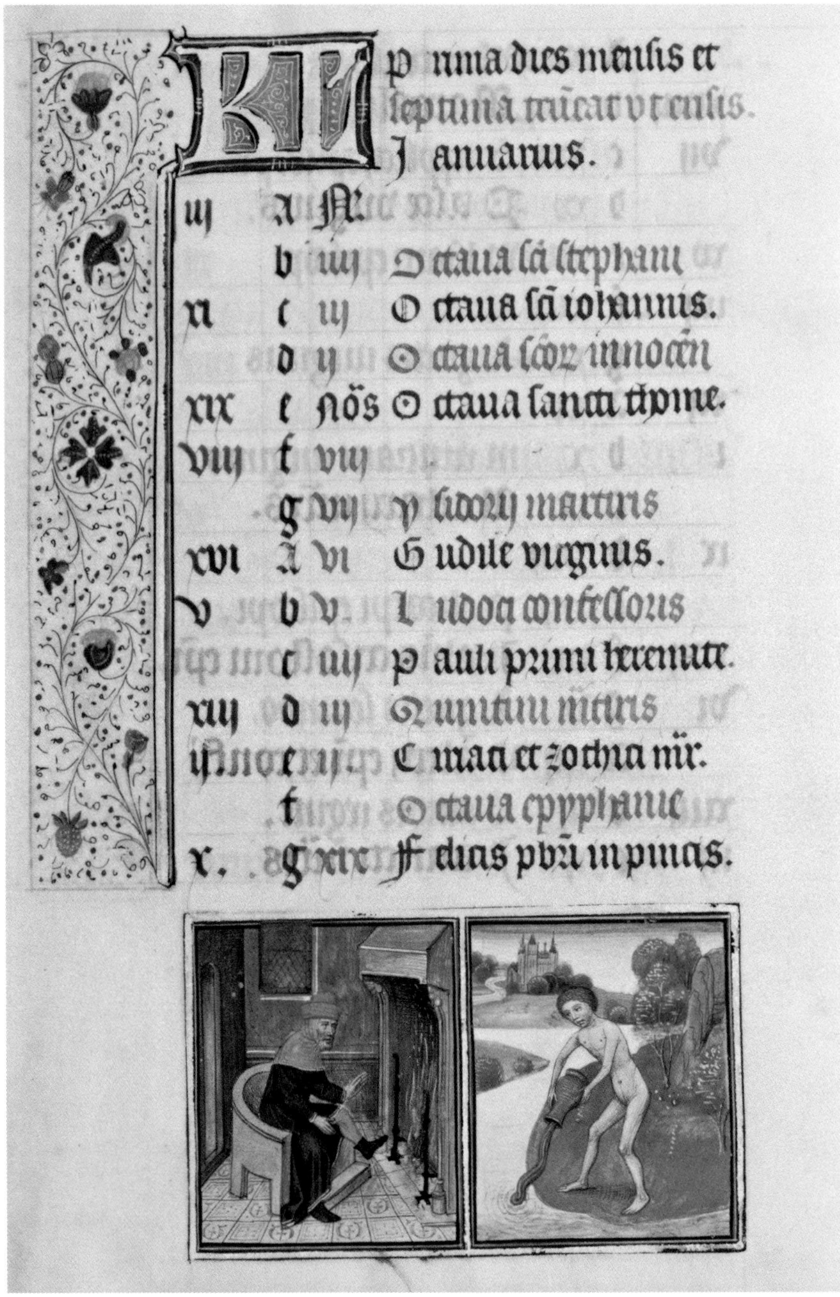

1 William Vrelant or workshop, *A Man Warming by a Fire; Zodiacal Sign of Aquarius*, Arenberg Hours, early 1460s, folio 1, tempera, gold leaf and ink on parchment bound between wood boards covered with purple velvet.

from a manuscript, and shows a man sitting on a semi-circular chair, leaning slightly forward in front of a hearth where a dullish, yellow fire burns; alongside this there is another picture of a nude adolescent out of doors pouring water from a jug into a river (illus. 1). I glance up from the manuscript and gaze through the large windows at the surrounding hills, and think about how regularly fires sweep through this area. I think too about the fires that habitually ravage Australia where I have lived for some 30 years: arriving unannounced, at times, they can be singularly destructive. Los Angeles, however, preoccupies me. In October 2003, the year before this introduction was written, wildfires advanced on three fronts, one of them pushing into the heavily populated neighbourhoods of northwest Los Angeles and the San Fernando Valley. A four-day toll left more than 2,000 homes destroyed, fifteen dead and some 500,000 acres scorched. Over 10,000 firefighters battled the flames.[1]

Thoughts on the grim effects of fire bring me back to fire's representation on the parchment in front of me. Called *A Man Warming by a Fire*, the image shows a handsomely dressed figure in a long, blue tunic and wearing a hat; his left leg is exposed to the hearth; a simple scene of simple, domestic pleasure. No major conflagration, no loss of life or destruction of property. The adjacent image, as I have said, depicts a naked youth out in the open, standing by a river into which he pours water. It's called *Zodiacal Sign of Aquarius*. Contrasting pictures, indeed: inside and outside; private and public; fire and its nemesis, water. One element used to combat the other; and both equally mercurial, something that the Australian artist Fred Cress (1938–2009) captures in his beguiling painting, *Two Waves,* where the two elements rub shoulders and, paradoxically, seem at ease in each other's company (illus. 2).

Moving on from Vrelant's solitary man in front of his hearth (and it is *his*: his home, his fire, his pleasure) to another image from a manuscript, *The Expulsion from Paradise*, of around 1360–70 (illus. 3), a work based on Genesis 3:24:

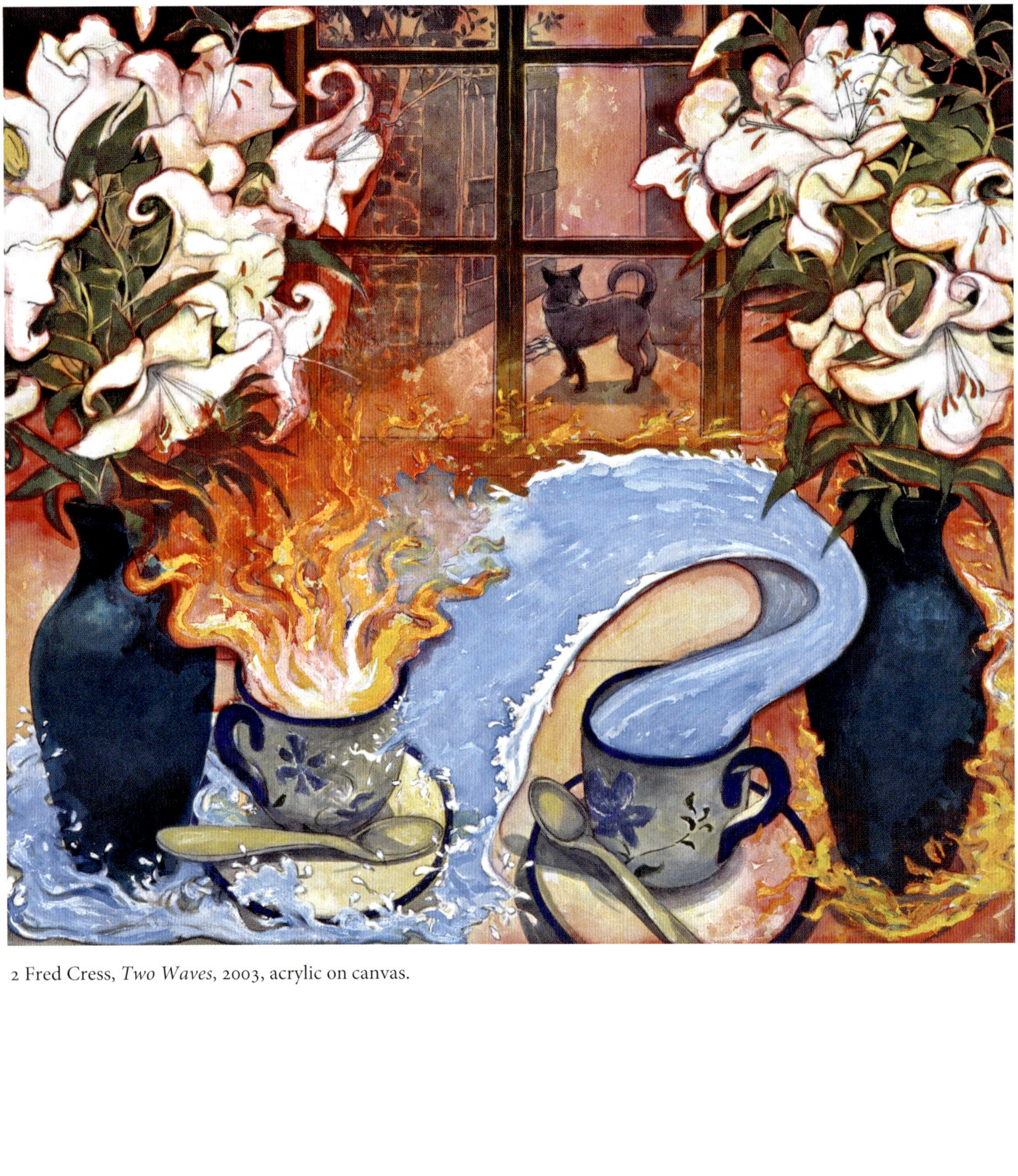

2 Fred Cress, *Two Waves*, 2003, acrylic on canvas.

> And the Lord God sent [Adam] out of the paradise of pleasure, to till the earth from which he was taken . . . and placed before the paradise of pleasure cherubim, and a flaming sword, turning every way.[2]

The domestic delights of fire (in the Vrelant) here give way to its damnatory potential. Recognizing this, the artist has shifted the flames from a sword-carrying angel (whose gesture is directed at Adam and Eve) to Eden itself: twelve tongues of fire extend the length of the garden, their colours echoed in the decorative red background.

Slippages of language are revealing: 'tongues of fire', the phrase just used, collapse this metaphor with taste, something that is incongruous if it were not for fire-eaters who seemingly devour flames at will. Less whimsical, although equally suggestive, are William Sansom's allusions to taste and other senses in his essay 'The Naked Flame', published in *Fire and Water* (1942). An auxiliary firefighter during the Blitz, Sansom had worked in advertising before the war. His essay concludes with a lengthy personal account of fire taken from a journal he kept while in

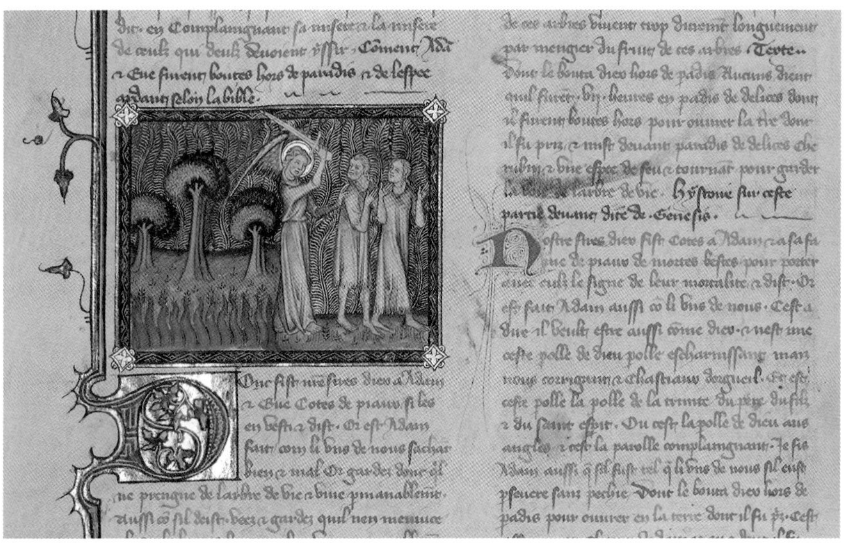

3 Attributed to the Master of Jean de Mandeville, *The Expulsion from Paradise*, c. 1360–70, tempera colours, gold leaf and gold paint on parchment.

service. The passages do not define fire, but rather evoke its impact on the senses. First, there is the 'smell of burning and disintegration. It is constant. It is solid.'[3] Then there is fire's impact on sight:

> The coppery red reflection of fire in the sky and on every building everywhere. This colour has the same terrible constancy as the smell of fire . . . It is a live colour, the colour of a living element . . .[4]

Aesthetics is soon introduced: 'From a distance, serving its part in the blitz panorama, it [fire] can be beautiful.'[5] Sansom does not invoke the 'Sublime' – a notion of a transcendental nature *in extremis* – but it is there, lurking in the background. Smell (and its inescapable association with the sense of taste) is revisited:

> I can remember the time a small scent factory caught alight. The heat of the fire burst a hundred bottles of concentrate essence and vaporised the perfume. That night we worked in an atmosphere drugged by the wildest, headiest smell . . . We drifted with a rosewater hose through a Persian garden where the trees were Mae West's hair and the walls the crystallized breath of every bosom since Zezebel.[6]

These are extraordinary descriptions, embellished in ways that bring to mind J.-K. Huysmans' *À Rebours* (1884), that quintessential Symbolist text where odours, colours and taste come together in an intoxicating mix.

Towards the end of his essay, Sansom writes: 'Finally, let us admit ourselves defeated by the naked flame. Let us concentrate instead upon its symptoms and results.'[7] Such an exercise, however, is not as straightforward as it seems. While it is easy to measure fire's deadly impact on lives and property, it is much more difficult to calculate its other outcomes and affects. For one thing, fire has its own vocabulary: its signifiers include flames, conflagration, charring, scorching, toasting, warming and blackening, and each signifier, in its own way, suggests different contexts and different meanings. In the wire reports about the Los Angeles fires summarized earlier, officials were said to be especially concerned about 'crowning', 'in which flames leap from one treetop to another, leaving firefighters on the ground all but powerless to stop them'.[8] It is tempting to think of the English artist Cornelia Parker, and her *Hanging Fire*, a work installed along the wondrous sculpture trail in the Forest of Dean in Gloucestershire (illus. 4). Consisting of a crown of flames made of cast iron, now heavily rusted, it hangs upside down high up in a ring of sycamore trees. The crown encircles one of these trees. Commenting on this sculpture, Rupert Martin writes:

> Drawing on the symbolism of the crown, the work relates to the royal nature of the forest, and the idea of the fall of kings. It also resembles a crown of thorns while the flames falling from above recall the Holy Spirit at Pentecost, their rust colour reminding one of the leaves falling in autumn.[9]

To pursue this analogy with Christ's Crown of Thorns, it could be said that this association locates fire in the redemptive. And this is not entirely fanciful. Take one of many examples, the Great Fire of London that raged for four days in September 1666 and destroyed over 13,000 houses, 87 churches and 44 of the City of London's livery halls.[10] Surprisingly, very few citizens were killed; what followed was a swift rebuilding programme that hinted at a new London. 'Hinted', because innovative designs were rejected, and much of the city's medieval, maze-like structure was retained.[11] Still, like the Phoenix, that legendary Egyptian bird who is said to live for 500 years and become young again after death by fire, a new London began to emerge from its own ashes.

London's burning and that of other great cities such as Chicago (1871) and Dresden (1945), which I discuss in chapter One, are public

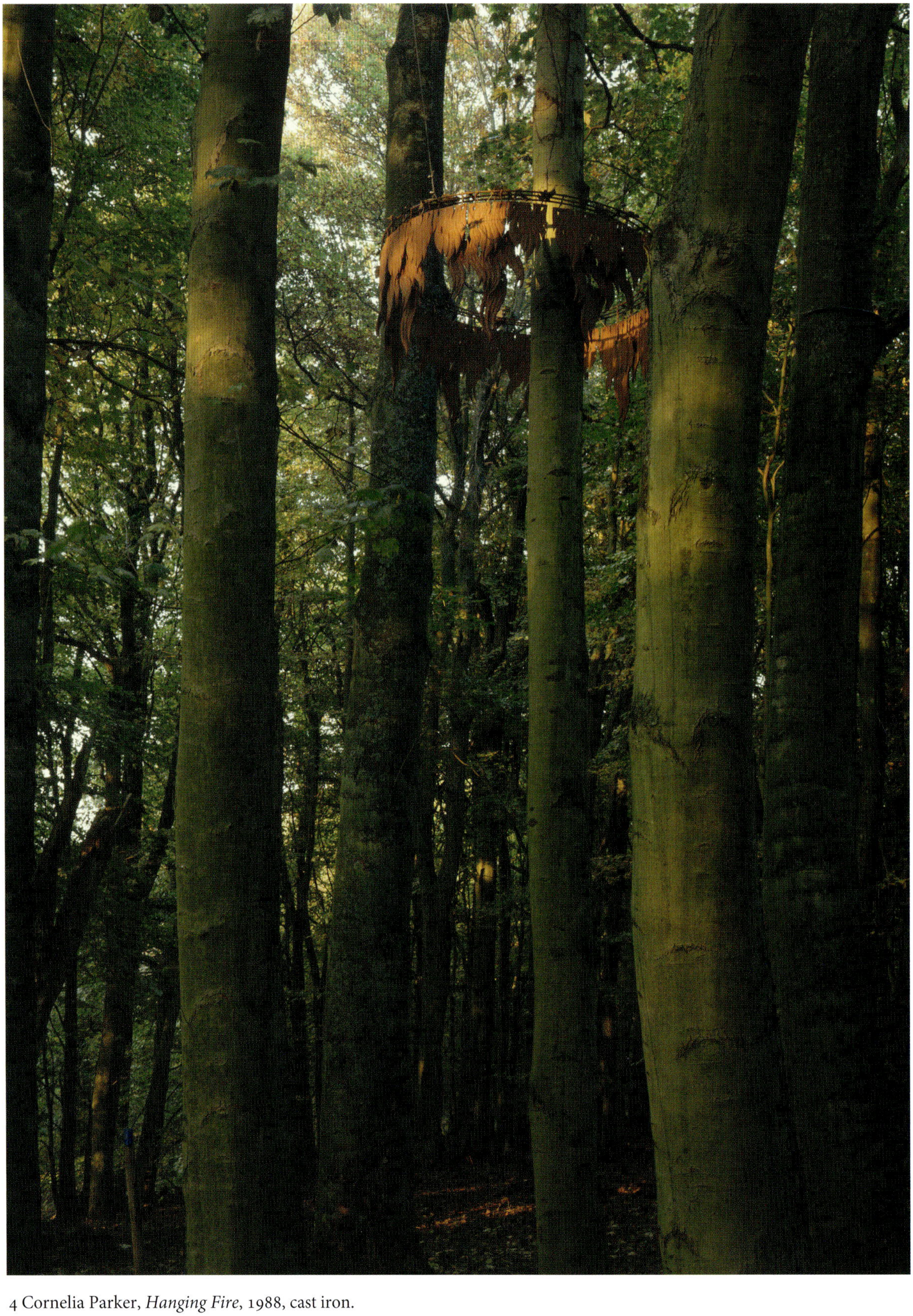

4 Cornelia Parker, *Hanging Fire*, 1988, cast iron.

fires whose imagery and related commentary have little to do with the inviting nature of the hearth, the campfire or the flame of a candle; little fires that speak of private spaces and special moments that along with other intimate flames I consider in chapter Three. Fire's nuanced and contrasting relationship to myth and religion, the focus of chapter Two, shows how fire may illuminate both literally and figuratively yet may also be understood as an element of dread and damnation. The fourth and final chapter looks at modern and postmodern artists who explore the full range of fire's mercurial and conflicting nature.

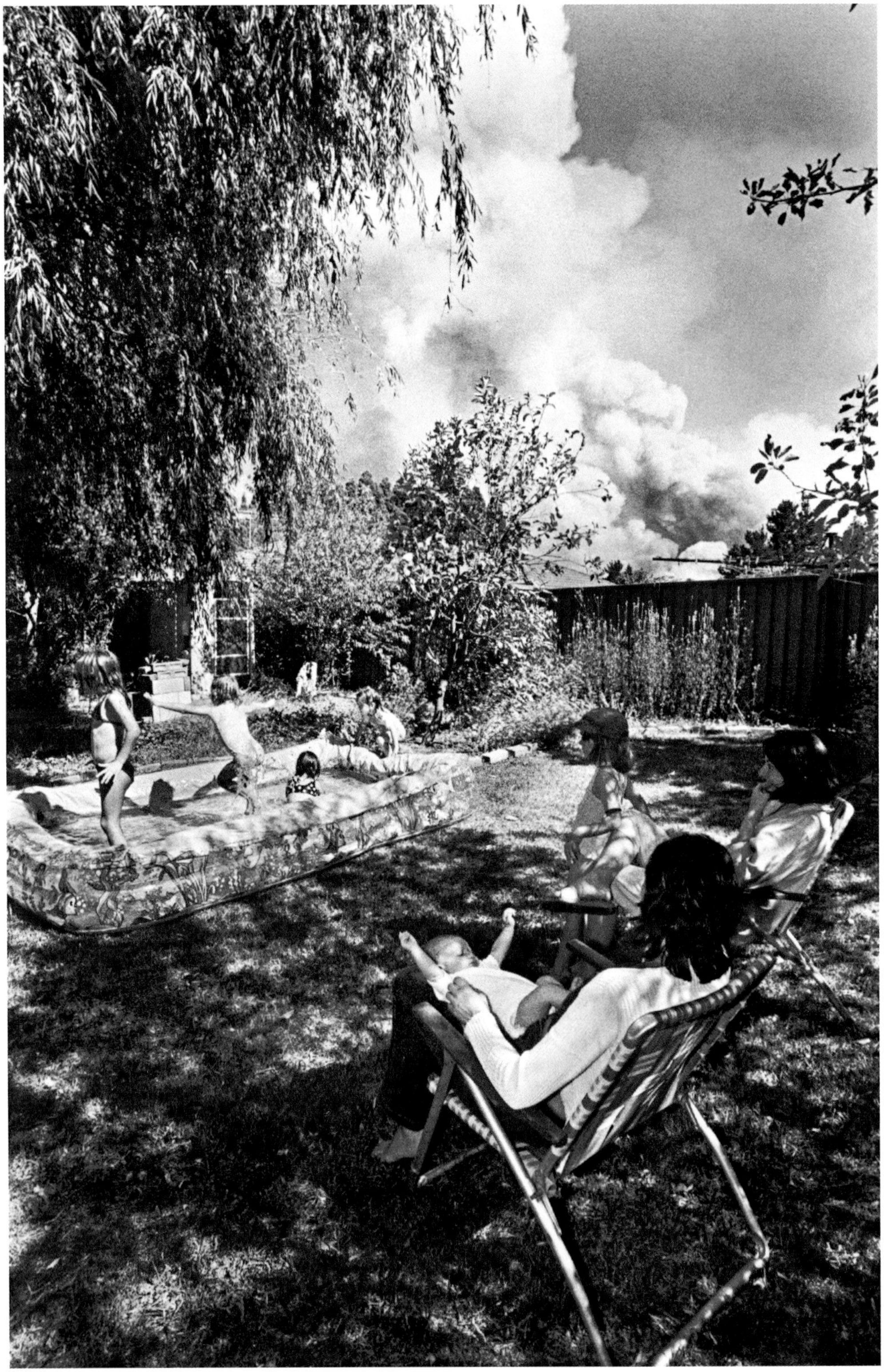

5 Philip Gostelow, *Backyard, Woodford (NSW – Black Christmas Bush Fires 2001/02)*, 2002, photograph.

one

Fabled Fires

What will doubtless prove the most destructive conflagration of modern times has been raging for two days and nights in our sister city of Chicago . . . A hundred thousand are homeless . . . It is the GREATEST FIRE IN MODERN HISTORY . . . We do not except even THE GREAT FIRE OF LONDON.

St Louis Times (11 October 1871)[1]

There are fires that enter the social imagination through various combinations of cause, effect and discourse and become 'Great Fires': London in 1666 and Chicago in 1871 immediately come to mind. A large medieval maze, London stood in contrast to Chicago, a distinctly modern city with cast-iron buildings, railroads and industries. The fires that swept through these two cities, however, had this in common: both eclipsed their origins in circumstances that may best be described as commonplace; and both went on to draw out responses, literary and visual, that became as celebrated as the fires themselves. Giving form to fire's contradictory nature, these responses shift easily from evocations of the beautiful into the terrible, a segue that sits comfortably with the understanding (which is nuanced yet slippery) of how such opposing affects cohabit in the 'Sublime', a term widely used in the nineteenth century and for at least two centuries before that.[2] This combination of fear and fascination is something that is always simmering in the reception of public fires; and it is one that distinguishes the so-called Great Fire of Edinburgh in 1824, an overlooked but galvanizing phenomenon in Scotland's early history. Civilian conflagrations, these Great Fires contrast with those dreadful firestorms that swept through Dresden and London during the Second World War, two other cities that feature prominently in the litany of legendary fires. Their blazes were wilful and bloody-minded, however, and this is reflected in the rhetoric surrounding them. All these fires were urban; by contrast, those that raged through New South Wales, Australia, in 2002, the 'most flammable country on earth',[3] took place, and continue relentlessly to do so, in that ill-defined zone between city and country.

Sydney: 'The Monster Arrives Early'

Sydney was caught by surprise on 8 October 2002. The dreaded summer of bushfires arrived early that year. It was an unusually hot day with low humidity and disturbing winds. Eighty fires criss-crossed the state of New South Wales that night. In front-page coverage of the fires the following day, the *Sydney Morning Herald* illustrated its report with a large, aerial photograph of burning houses in Thurlgona Road, Engadine, an outer Sydney suburb (illus. 6). Staff reporters wrote:

> Ten homes in Engadine went up in flames, but there were also blazes in Llandilo, Richmond and Windsor-Castlereagh . . . A major blaze was also burning near Bilpin on Bells Line of Road in the Blue Mountains National Park last night, while Rural Fire Service volunteers were fighting two fires that had cut the Newell Highway near Narrabri. In all, 10 houses had been destroyed in Engadine last night.[4]

Simple and to the point, these matter-of-fact words belie their dramatic headline, 'The monster arrives early', a hugely evocative phrase that introduces an imaginative dimension (as always) into the discourse on fire. Some seven years later, on 7 February 2009, the Black Saturday bushfires in the state of Victoria would attract similar headlines. 'Hell's Fury', announced the *Daily Telegraph*, reporting on the 96 dead (and rising) and the 750 homes destroyed.[5] Black Saturday would finally

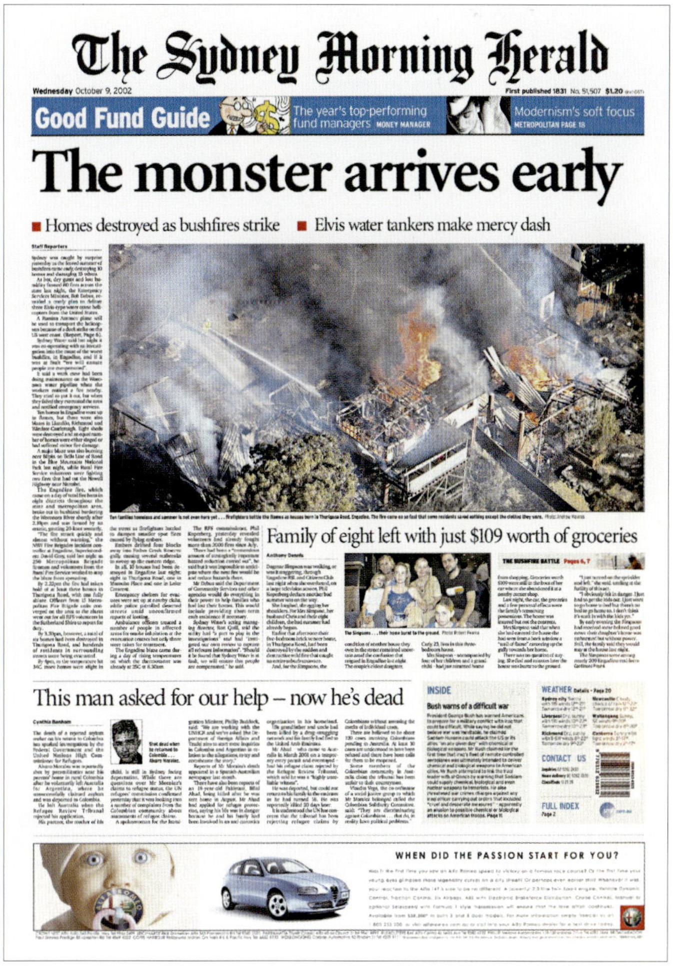
The Sydney Morning Herald

Wednesday October 9, 2002

Good Fund Guide

The year's top-performing fund managers

Modernism's soft focus

The monster arrives early

■ Homes destroyed as bushfires strike ■ Elvis water tankers make mercy dash

Family of eight left with just $109 worth of groceries

This man asked for our help – now he's dead

INSIDE

CONTACT US

FULL INDEX

WHEN DID THE PASSION START FOR YOU?

6 'The monster arrives early', *Sydney Morning Herald*, 9 October 2002.

result in 173 deaths, the highest ever mortality rate from a bushfire in Australia.[6]

In New South Wales in 2002, surprisingly, no lives were lost but over 570 buildings and over 730,000 hectares (1.8 million acres) of land were destroyed.[7] By way of contrast to the alarming headline discussed above, 'The monster arrives early', Darren Pateman's photograph taken in the small town of Cessnock in the Hunter Valley of New South Wales (illus. 7) suggests something entirely different. Titled *The Ashes*, it is a wry reference to the celebrated cricket confrontations held roughly every other year between England and Australia. This epithet dates from the Australian tour to England in 1882–3, when a surprising loss by England in a low-scoring game led to a tongue-in-cheek advertisement in the *Sporting Times* declaring the imminent cremation of the 'body' of English cricket (with the 'ashes' to be taken to Australia). The ashes, as it were,

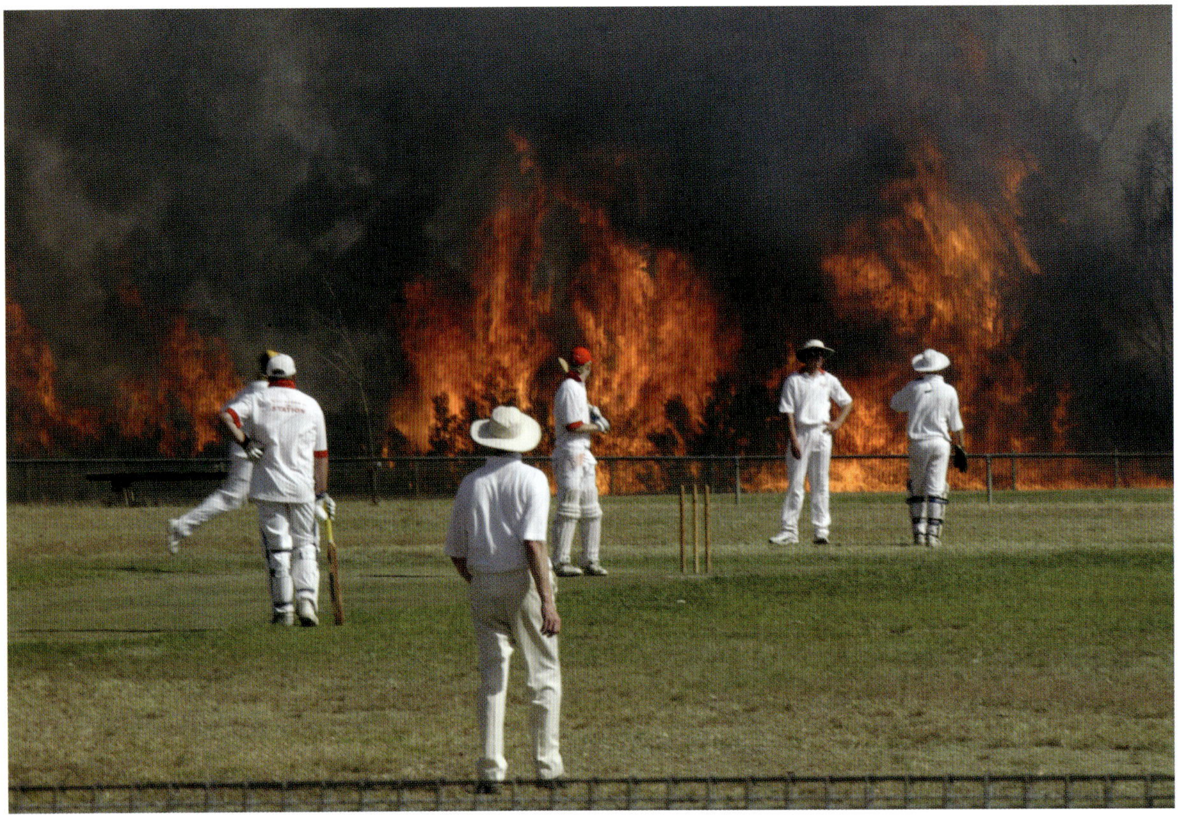

7 'Cricket carries on regardless of huge fires in Cessnock', alternatively titled 'The Ashes', *Sydney Morning Herald*, 19 October 2002; colour digital print, NCH picture by Darren Pateman.

are contained in a small terracotta urn of unspecified origin whose contents are believed to be variously a stump, the outer casing of a cricket ball, a veil or most likely a ball itself.[8] These are fripperies, of course; but they do illustrate the floating meanings of fire and their appropriation into popular culture. Pateman, for one, recognizes this. He shows white-flannelled cricketers continuing their game as huge fires rage beyond the boundary fence, but he also hints at other agendas. The photograph is all about bravado and recklessness and what, in Australian parlance, might best be described as 'larrikinism' – a legendary irreverence in the face of officialdom.

Another photograph of a bushfire in New South Wales, by Philip Gostelow, shows a blaze in Woodford (in the Blue Mountains, west of Sydney); taken some months earlier, it presents a similar scenario (illus. 5). A mother sits contentedly on an outdoor chair, baby on lap, while children frolic in a plastic pool. All this takes place in backyard suburbia; meanwhile, ominous clouds of smoke swell in the background. Gostelow's comments on this work capture the fire's incongruous presence:

> I'd set-up [sic] the inflatable pool in [the] b' garden – taking a diversionary respite from the anxiety and constant threat of giant bushfires approaching the B Mtn [Blue Mountains] townships. It was mid afternoon, we invited some neighbors over for their kids to have a paddle – to try take our minds off the bushfires. The scene was bizarre. A seemingly normal backyard in the summer, and then you'd look up and see the specter of fire in the southern sky. The buffeting noise of water tanker helicopters overhead, fire truck sirens echoing down the highway. It all made for a surreal atmosphere and a mood of suppressed anxiety. And a sense of anticipation.[9]

The Blitz, and the Heroizing of the Firefighter

The 'monsters' in Pateman's and Gostelow's photographs are there but their presence is muted. Not so in a series of paintings titled *Menace* by Paul Dessau (1909–1999), a member of the London Auxiliary Fire Service (AFS) during the Second World War, and a founder member of the Firemen Artists' Committee.[10] Established to promote the work of 'firemen artists', the Committee was soon recognized by the Home Office; an official circular of February 1940 said that, 'where circumstances permitted', artists should be allowed to contribute to the war effort in ways they saw fit.[11] Between 1941 and 1944, these 'firemen

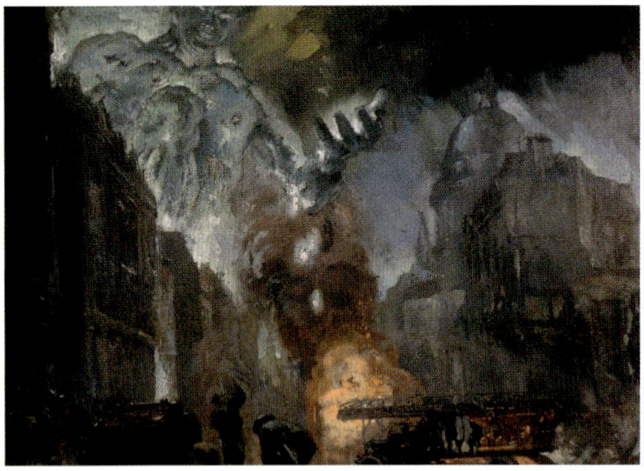
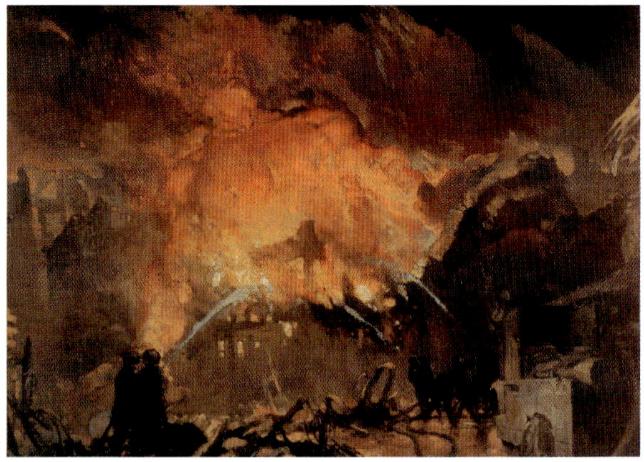

8, 9 Paul Dessau, 'Overture', Part 1 of *Menace* series (left), and 'Crescendo', Part 2 of *Menace* series, 1941, oil on canvas.

artists' held four exhibitions at the Royal Academy; two other shows travelled to the USA and Canada in 1941.[12]

Like other 'firemen artists', including Leonard Rosoman (*b.* 1913; see below) and Norman Hepple (1908–1994), Dessau had personal experience of the Blitz. Heralded on 13 August 1940 ('The Day of the Eagle') with a massive Luftwaffe attack on air bases and aircraft factories in the southern counties, the German onslaught soon moved to London itself. The date 7 September was especially terrible. Commencing in the late afternoon, 300 German bombers, escorted by 600 fighters, arrived in two waves over the Docklands area in the East End. Many of the bombs, however, fell on the residential areas around them.

> The bombers, seeking the docks, dropped their bombs on some of the poorest and most overcrowded streets of London, their slum buildings and tenement houses more vulnerable than most to the pounding blasts of bombs and the ensuing fires.[13]

As a result 5,448 Londoners were killed. At the height of the Blitz, London was pounded for 57 consecutive nights between September and November 1940; isolated major raids continued into 1942.

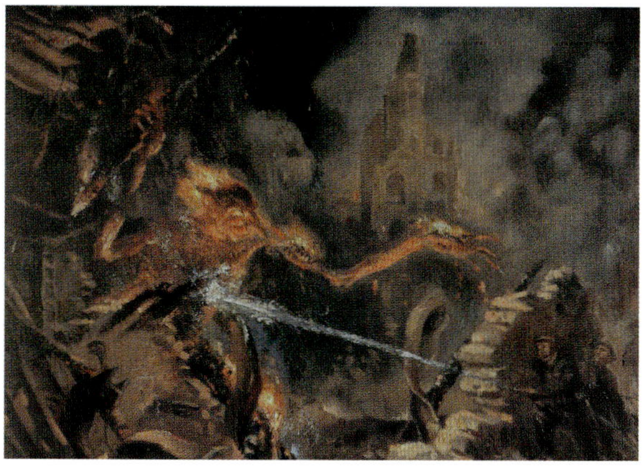

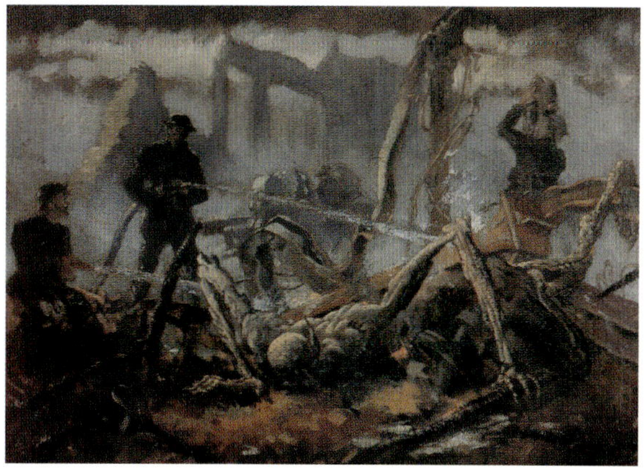

10, 11 Paul Dessau, 'Rallentando', Part 3 of *Menace* series (left), 'Diminuendo', Part 4 of *Menace* series, 1941, oil on canvas.

It is no surprise, then, that Dessau takes recourse to the fantastical to represent this devastation. *Menace* consists of four canvases respectively called 'Overture' (illus. 8), 'Crescendo' (illus. 9), 'Rallentando' (illus. 10) and 'Diminuendo' (illus. 11). (These titles, incidentally, reflect the artist's lifelong passion for music.) Dessau depicts the fire (in some urban setting amid tall buildings) as a colossal, grey-blue giant towering above the architecture. Bearing a devilish smile, the giant beckons to the firefighters that are dwarfed below ('Overture'); in 'Crescendo', he is transformed into a red-hot creature, at whom the firefighters in the foreground direct their hoses; by part three, 'Rallentando', the muscular monster is reduced to a skeletal, fiery demon, carrying a burning, spear-like instrument in his right hand; in the fourth and final picture, 'Diminuendo', the creature is tamed, now simply a fallen, bony figure, face down in muck and debris. In contrast to the first three canvases, the firefighters appear calm and collected; victory is theirs!

Dessau's incredible imagery addresses the horrors of fire, but it is equally about the heroism (and mythicizing) of the firefighter. Such adulation was not new. Open flames in nineteenth-century urban America were a constant threat; and volunteer fire departments did their best to combat them. One of these volunteers was Nathaniel Currier (1813–1888) who, together with James Merritt Ives (1824–1895), set up the long-standing printmaking firm Currier & Ives, which published a multitude of highly popular series, including 'The Life of a Fireman' (1854–66).[14] Making much of the ingredients of modernity – steam-powered pumps replacing hand-pumps, steam engines taking over from double-decker hand-pumpers – these images, painted by staff artist Louis Maurer (1832–1932), carried such stirring captions as 'Jump her boys jump her' and 'Now then with a will – Shake her up boys'. This gendering of fire as 'female' adopted a different gloss in Thomas Edison's groundbreaking silent documentary-drama, *Life of An American Fireman* (dir. Edwin S. Porter, 1903), staged with the help of New Jersey's Newark and Orange fire departments. Here firemen break down doors and windows, a woman screams, and a child is carried down a ladder, a scenario that had been anticipated by the Victorian artist John Everett Millais (1829–1896) in *The Rescue* (illus. 12).[15] His first painting of a contemporary subject, it shows a sturdy fireman emerging from a burning building with two children: one is caught between his arms, the other, a young boy, clutches his back; the kneeling mother in her nightdress stretches out her arms with theatrical poignancy in a welcoming gesture. If here the firefighter is represented both as saviour *and* father, then in the Hollywood action movie *Backdraft* (dir. Ron Howard, 1991), which was produced with the full cooperation and technical support of the Chicago Fire Department, male bonding and braggadocio dominate. *Backdraft* treats fire as an alien but living female being; and the sentiments are openly misogynistic: 'let's hit this bitch head on . . .'; 'she's hot and smoky'. More recently, *Ladder 49* (dir. Jay Russell, 2004), continued this glorification of the firefighter and his homo-social environment, but without the obvious sexualizing of *Backdraft*. Parting company with the generally enthusiastic reception of *Ladder 49*, Manohla Dargis in the *New York Times* wrote (correctly, I believe) that *Ladder 49* is

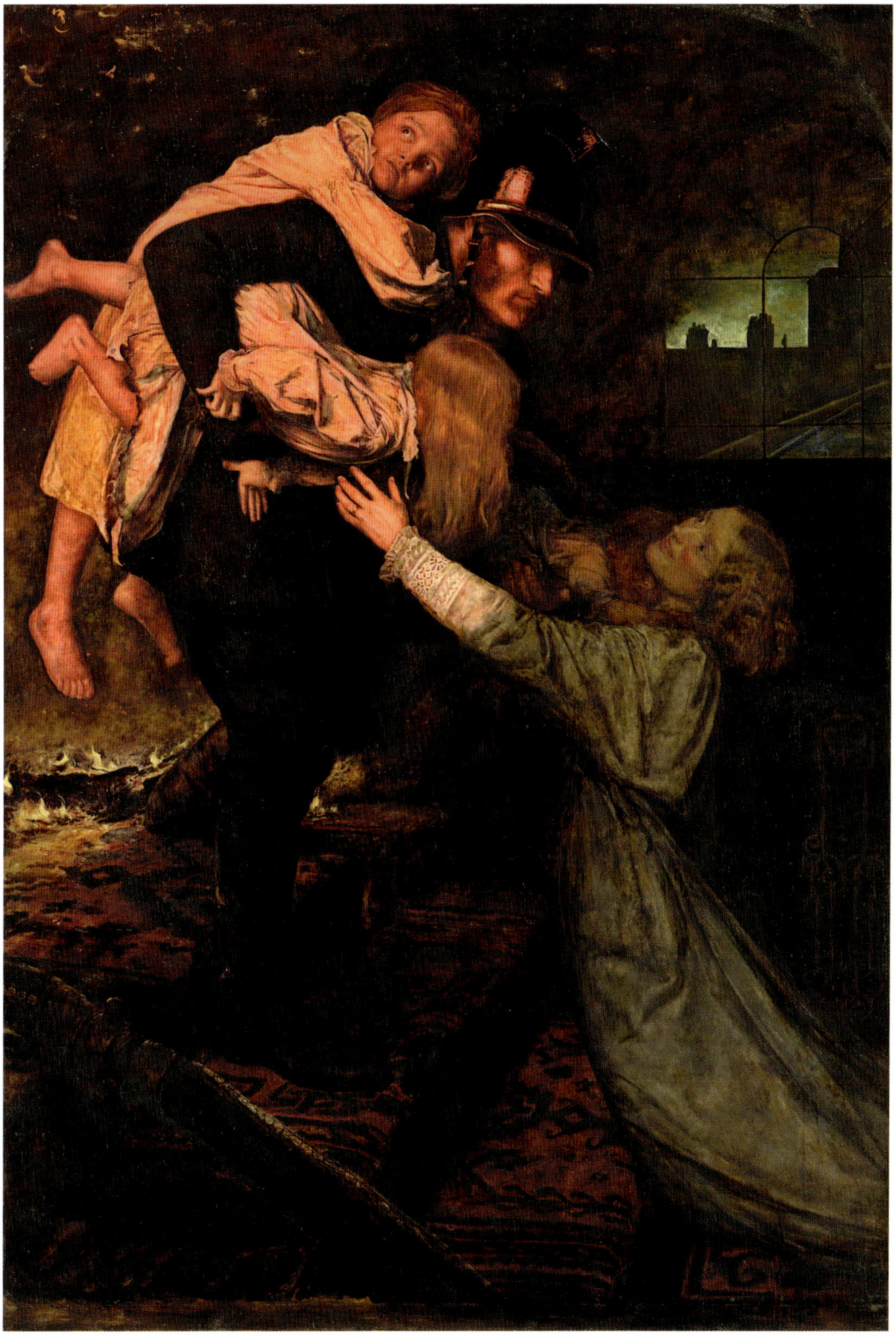

12 John Everett Millais, *The Rescue*, 1855, oil on canvas.

> essentially a male weepie about strong, simple men and the strong, simple women behind them, and as such it's platitudinous nonsense. What makes this nonsense more galling than usual is that while Ladder 49 might have started out as a heartfelt attempt to honour those in the line of literal fire, it weighs in as an attempt to exploit their post-Sept. 11 symbolism.[16]

By contrast, the 54-minute documentary by Mike Lennon, *Brothers . . . On Holy Ground* (2002), notwithstanding its sensationalist title, is a hugely affecting, tightly edited film about the firefighters of 9/11, 353 of whom died at the site of the World Trade Center. A firefighter himself and independent filmmaker, Lennon began filming and interviewing two weeks after searching for survivors. Speaking to colleagues at firehouses that had lost men, these men's surviving wives and residents of lower Manhattan, Lennon tells the story of a close-knit band of men with an intense sense of duty and comradeship. 'I don't feel like a hero', says firefighter Rick Farro, while John Santore, of Ladder Company 5, whom Lennon had interviewed in 1999 but who would die on 9/11, simply says: 'You might get killed – but you do it.'

Back to London and the Blitz: the heroic is implied in Leonard Rosoman's *A House Collapsing on Two Firemen, Shoe Lane, London, EC4* (illus. 13), but the wall that crashes down on the two abruptly painted firefighters denies this. Debris and embers fly everywhere. It's a dramatic picture, all noise, movement and colour. Unlike Dessau's *Menace*, discussed above, with its sci-fi imagery, it draws attention to the specific and the 'real': Shoe Lane, London, EC4. Rosoman had trained at the Royal Academy Schools in London, and at the Central School of Arts and Crafts; he was also, as mentioned before, a firefighter in the AFS. William Sansom's essay 'The Wall', from *Fire and Water*, the anthology cited in the Preface of this book, may bear on Rosoman's picture. It chronicles the 'third job that night' undertaken by Sansom and three other firefighters, one of whom, named 'Len', may have been Rosoman himself:

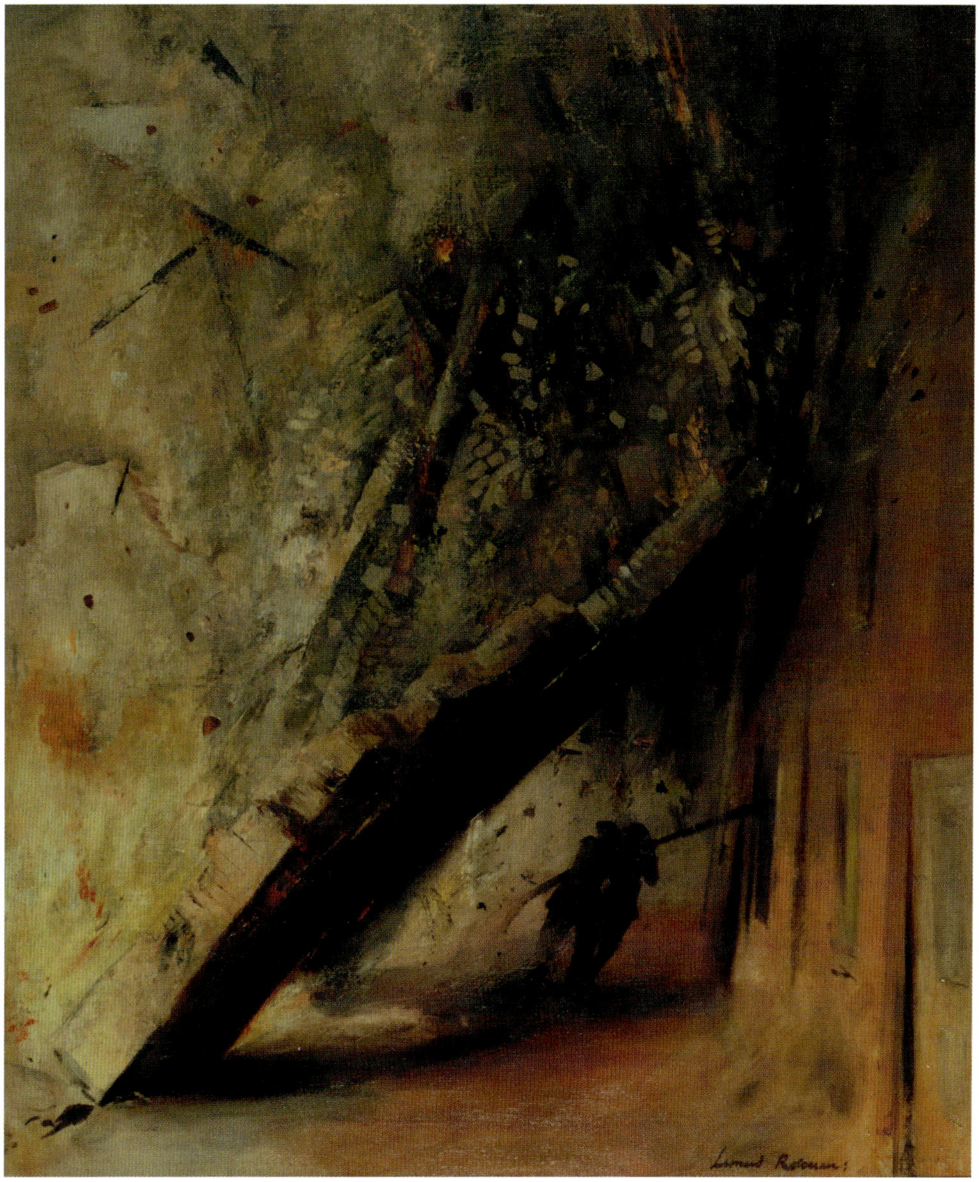

13 Leonard Rosoman, *A House Collapsing on Two Firemen, Shoe Lane, London* EC4, 1940, oil on canvas.

> The 'job' involved playing a fifty-foot jet up the face of a tall City warehouse . . . You know the fire will fester for hours yet. However, that night, the blank, indefinite hours of waiting were sharply interrupted by an unusual sound . . . a . . . rattling crack of bursting brick and mortar . . . And then the upper half of that five storey-building heaved towards us . . . A wall will fall in many ways. It may sway over to one side . . . It may crumble . . . It may remain intact and fall flat. This wall fell as flat as a pancake.[17]

This is an apt description of the canvas that shows the wall, inclined at a sharp angle towards the ground, falling like a two-dimensional toy soldier.

An elderly Leonard Rosoman appeared in a BBC television documentary, *The Blitz: London's Firestorm* (dir. Louise Osmond, 2005), about the devastating German aerial onslaught on London on 29 December 1940 that saw over 10,000 incendiary bombs fall on the city in the first two hours of the raid, and which has been compared in its effects to London's Great Fire of 1666. Rosoman's comments in this film confirm that he was indeed with Sansom in Shoe Lane, a small alleyway west of St Paul's Cathedral, on that fateful night. The interior of a timber Victorian book warehouse was ablaze, and only the brick wall kept it standing. Describing the eerie sound that the soon-to-collapse structure was making, Rosoman says: 'It was in fact all this red-hot brick, all five storeys of it on the point of combustion . . . [one felt] absolutely, absolutely helpless.'

Samuel Pepys and the Great Fire of London 1666

> What did Pepys feel when he saw the Great Fire of 1666? – It was just unbelievable.[18]

Here the 'unbelievable' refers to the firestorm, mentioned above, that raged through the heart of London four days after Christmas in 1940. No need, however, to speculate on what the diarist Samuel Pepys (1663–1703) felt when he first laid eyes on London's conflagration – it was indifference: having just woken in the early hours of that fated morning he saw the fire from a distance but was not really concerned. This is unlike the other famous diarist of 1666, John Evelyn, whose journal entry for 2 September, the day the fire began, while similarly brief, noted how 'exceedingly astonished' the writer was.[19]

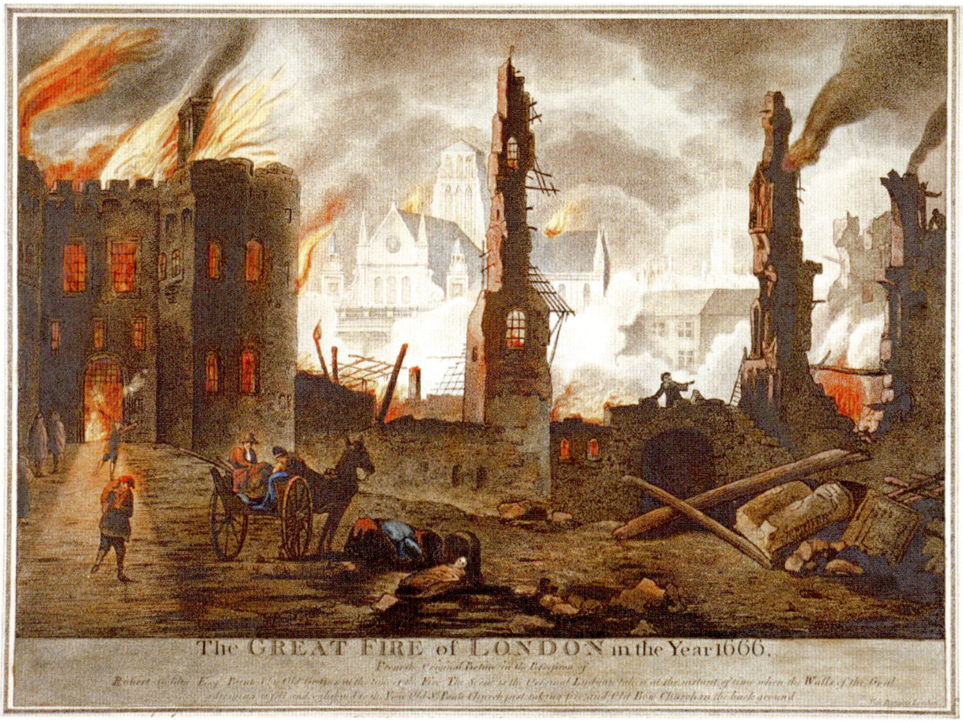

14 *The Great Fire of London in the Year 1666*, 1792, engraving after a coloured aquatint by Jan Griffier the Elder, *London on Fire*.

On that morning, a Sunday, what would become known as the Great Fire of London began surreptitiously (illus. 14). The history of that conflagration continues to be written; combining fact and anecdote, myth and mythmaking, it is a construction of fire that reverberates in the imagination.[20] It even appears in a great many childrens' books, which is no surprise, really, since bonfires and fireworks and things that go 'crack' and 'pop' in the night – fear and fun – are so much part of childhood experience. Karen Wallace's recent *London's Burning!* builds on the well-known and much garnished story of the fire's source in the house of the king's baker, Thomas Farriner (sometimes referred to as Farynor), a contractor providing biscuits to the Navy, in the (now) eponymous Pudding Lane. Wallace's tale focuses on the young Harriet Farynor, her dog Ginger, her friend Nelly, and their respective experiences of the fire. It was while Harriet and Nelly 'were asleep in the attic bedroom at Aunt Gwen's house', that Harriet awakened suddenly. She ran to the window, looked out, and saw that

> the sky was lit up with a dull orange glow. As she watched, the rustling noise turned into a roar and a man came running down the street. She lifted the catch on the window and pushed it open.[21]

London is burning: only the boats in the Thames and the buildings in the immediate foreground escape the fury of the flames (illus. 15). Harriet looks out, wide-eyed and distraught.

Samuel Pepys was similarly awoken very early on that fateful morning and he too went to a window to see what all the commotion was about:

> 2nd. Lord's Day. Some of our maids sitting up late last night to get things ready against our feast today, Jane called us about 3 in the morning, to tell us of a great fire they saw in the City. So I rose, and slipped on my nightgown and went to her window; and thought it to be on the back side of Markelane at the furthest; but being unused to such fires as fallowed [sic], I thought it far enough off, and so went to bed again and to sleep.[22]

This image of Pepys dressed in his nightgown and concluding that the fire is not menacing after all, borders on bathos. Like the little Harriet, also just emerged from slumber, his view of the fire is one framed, picture-like, in a window. Observed from a distance, fire always is a silent, mesmerizing spectacle: all colour and shifting configurations of flame and smoke; nothing is heard, let alone felt. A contemporary Dutch engraving, *London Verbrandt* (*London on Fire*), shows well-heeled Londoners, some with their pet dogs, looking at the distant city ablaze.

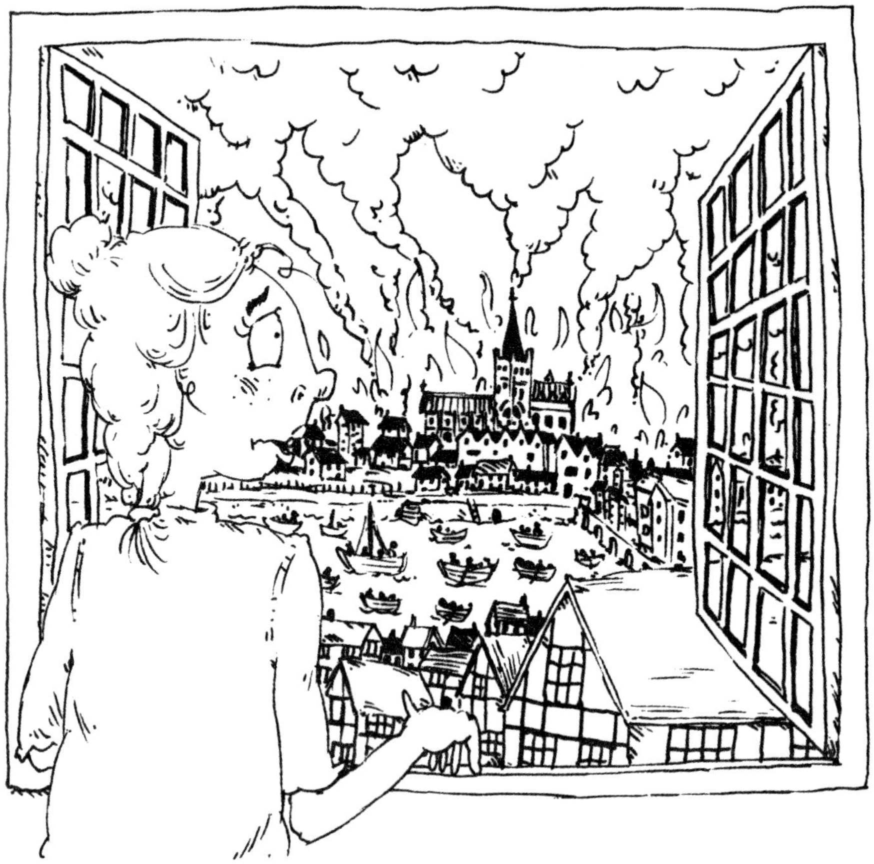

15 An illustration by Jamie Smith from *London's Burning* by Karen Wallace (London, 1997).

To all intents and purposes, they could simply be taking their pleasures outdoors. (This is not as fanciful as it might seem: some years ago, in high summer, I was with friends enjoying a leisurely lunch in a small restaurant near Sydney overlooking the Hawkesbury River: half a mile or so across the water we could see fires engulfing homes and bush. Our meal continued.)

Pepys went back to sleep and rose a few hours later, at about seven o'clock, to see how the fire had progressed. What reveries occupied Pepys as he slept, we will never know. At any rate, when, for the second time, he 'looked out at the window [he] saw the fire not so much as it was, and further off'.[23] The writer Stephen Porter takes up the story:

> Only when Jane [the maid] came to tell him that she had heard that more than 300 houses had been burnt down did he realise the seriousness of the situation. Although the figure could be no more than hearsay, it certainly indicated that the fire was an unusually large and threatening one.[24]

And so it proved. The prevailing winds sent the blaze westerly through the compact buildings along Thames Street to the west of Pudding Lane. Come the end of Sunday, the fire

> had burned along the riverfront for about half a mile, and had engulfed the area contained within an arc running from the river close to Queenhithe through the Bush Lane area to the top of St Michael's Lane and Fish Street Hill at Cannon Street, the northernmost point reached by this time.[25]

That night, after Pepys had dined (the niceties of daily convention were always to be observed), he went

> to White-hall by appointment, and there walked to St James's Park, and there met my wife and Creed and Wood and his wife and walked to my boat, and there upon the water again, and to the fire up and down, it still increasing and the wind great. So near the fire as we could for smoke; and all over the Thames, with one's face in the wind you were almost *burned with a shower of Firedrops.*[26]

A hugely evocative description that invokes the gentle in an otherwise ghastly scene.

On Monday (the following day), another extremely hot day, the fire spread to a different geography and to a different constituency: it had burnt out the poor, those in Pudding Lane and its surrounding areas, and now it engulfed those in the City, in the more salubrious Lombard, Fenchurch and Gracechurch Streets.[27] Fire, if nothing else, is the great leveller. For another two days the fire raged. Tuesday, 4 September, was particularly dire: the Guildhall was damaged, while St Paul's Cathedral was destroyed; the fire then broke through the City walls into the area around Fleet Street. Pepys again:

> Only, now and then walking into the garden and saw how horridly the sky looks; all on fire in the night . . . and endeed it was extremely dreadful – for it looks just as it was at us, and the whole heavens on fire. I after supper walked into the dark down on Tower-street, and there saw it all on fire at the Trinity house on that side and the Dolphin tavern on this side, which was very near us – and the fire with extraordinary vehemence.[28]

As the fire progresses, and Pepys's direct experience of its ferocity increases, he moves from being an observer to a participant, although one who still has the savvy to bury his wine and prized 'Parmazan' cheese in a pit in his garden. Fire even in extremis can still elicit unexpected responses.[29]

The Great Fire devastated London. Remarkably, very few people were killed: six, according to all accounts. Loss of property was another matter. Somewhere in the region of 13,000 houses were destroyed and at least 65,000 people were made homeless; 87 parish churches, six chapels and 44 livery halls were gutted. These statistics vary according to different sources.[30]

In the aftermath of the Great Fire, there was a period of swift urban reconstruction, most of which was complete within five years. Various inventive designs for London were proposed. These ranged from the grid-like designs of Richard Newcourt, Captain Valentine Knight and Robert Hooke, to John Evelyn's squares and piazzas. In the end, Christopher Wren's (much modified) designs were accepted after a series of manoeuvrings that were invariably informed by political, religious and social matters.[31] London, after the Great Fire, still retained the architectural networks of an earlier, medieval time. Still, there was enough change to the urban landscape to suggest that a corner had been turned. Evelyn had anticipated as much when he wrote to Sir Samuel Tuke some three weeks after the fire, saying, 'Truly there was never a more glorious phoenix upon earth . . . if it do at last emerge out of the cinders.'[32] Wren certainly recognized this when he used the phoenix, that great symbol for resurrection through fire, as the main design feature for the pediment relief over the south portico of what would become his most enduring achievement: the new Cathedral of St Paul's (illus. 16). Accompanying Wren's phoenix is the motto from Matthew 27:63 – Resurgam: 'I will rise again.'

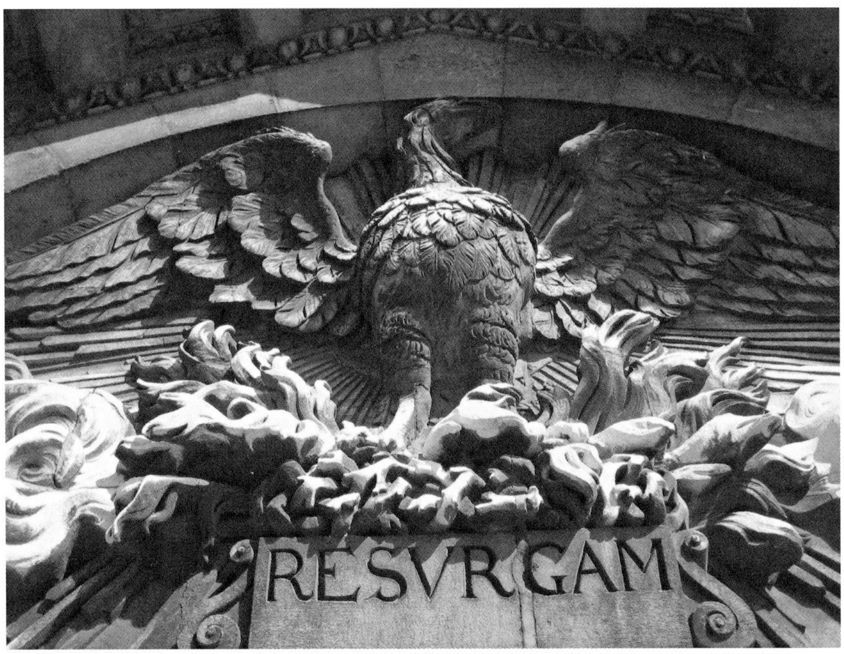

16 Phoenix, detail from pediment relief over south portico, St Paul's Cathedral, London.

A City and its Synagogue: Dresden Destroyed

In his recent study of 1666, Neil Hanson notes in passing that there are 'many close parallels between the devastation in London caused by the Great Fire and that in Dresden after the firestorm brought about by Allied bombing in 1945'.[33] This dispassionate aside falls beyond the scope of this study. What does concern me, however, are some of the terrible configurations of Dresden's history brought about by fire.

The bombing of Dresden began at 9.51 pm on Tuesday, 13 February 1945. Contrary to legend, Dresden was not an innocent city. The Nazis had established a strong presence even before Hitler came to power;[34] and the city contained manufacturing, communications and services crucial to Germany's war effort.[35] Whether this justified the dropping of more than 3,900 tons of high explosives and incendiary devices on the city is of course another matter entirely. What ensued was a horrific firestorm: at least 25,000 citizens perished and 39 square km (15 square miles) of the city's historic centre lay in ruins (illus. 17).[36]

> Bodies will be piled up in one of the main squares. They will be placed on huge slatted shutters, salvaged from the display windows of one of the city's department stores, then burned in the thousands to stop the spread of disease.[37]

First fire destroys, shockingly, and then it is employed to cleanse.

A Jewish slave labourer, Ben Helfgott, in a labour camp at Schlieben, later recalled the extraordinary red sky above Dresden, some 100 km (62 miles) away: 'Not only could we see it, the earth was shaking.'[38] Bombardier Miles Tripp, who was part of the second wave of attacks on the city, could also see the blaze from afar: 'Although we were forty miles from Dresden, fires were reddening the sky ahead.' Once over the target, he described how the city streets had been transformed into a

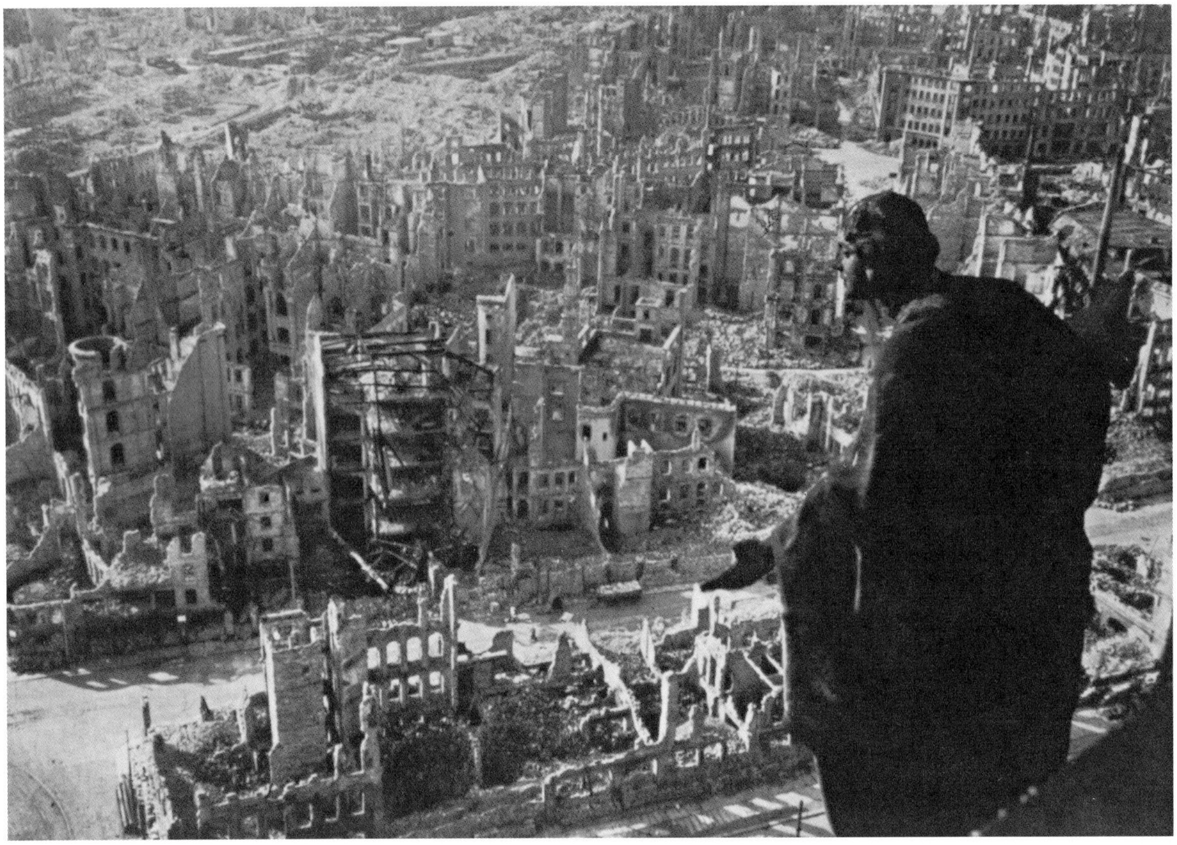

17 Walter Hahn, *View of the Dresden Ruins from the Roof of the new Town Hall*, 1945, photograph.

> fantastic latticework of fire. It was as though one was looking down at the fiery outlines of a crossword puzzle; blazing streets stretched from east to east [sic] from north to south, in a gigantic saturation of flame.[39]

And then, in terms that suggest the terrifying fascination of it all, he says, 'I was completely awed by the spectacle', words that form a common refrain in the distant (and mesmerizing) observation of great fires as seen from a distance.

No such balm for Kurt Vonnegut who, as a prisoner of war in Germany, had witnessed the destruction of Dresden by Allied bombers, an experience that inspired his novel *Slaughterhouse-Five* (1969). Through the mercurial character of Billy Pilgrim he describes the horror. Together with other American prisoners of war, Pilgrim was housed in a meat locker when the bombs began to fall. A guard would climb the stairs, peer out, and then report back to the other guards: 'There was a fire-storm out there. Dresden was one big flame. The one flame ate everything organic, everything that would burn.'[40] When Pilgrim and the Americans finally came out of the shelter, 'the sky was black with smoke . . . Dresden was like the moon now, nothing but minerals. The stones were hot. Everyone else in the neighborhood was dead.'[41] Then follows a scenario that deals with fire in an altogether contrasting circumstance. Arriving at a suburb that had escaped the devastation,[42] Billy, the guards and the other Americans come across an inn that is open for business. Inside all is snug and quiet. There is candlelight and three fireplaces. Dresden, meanwhile, was gone.

Let us return to Dresden in 1840, when it was still a city of culture (*Kulturstadt*) and of tolerance. The city's small but thriving Jewish community, determined to build a house of worship, commissioned the 35-year-old architect Gottfried Semper, a professor at the Dresden Academy of Arts. On 8 May 1840, the synagogue was consecrated.[43] Move forward to the night of 9 November 1938, the infamous *Kristallnacht* – the 'Night of Broken Glass' – when the Nazis vented their spleen on Jewish-owned businesses throughout Germany. Dresden was not exempt. Nor was Semper's synagogue, which Sturmabteilung (SA) men set alight. Otto Griebel, an artist, first saw the destruction of the synagogue from afar (like Pepys who, in very different circumstances, looked out at London burning in 1666). Unlike the diarist, however, Griebel immediately went to the site. There he witnessed firefighters, inactive, simply looking on at the conflagration.[44] Five years earlier, in 1933, soon after the Nazis came to power, SA troops had put fire to a different purpose:

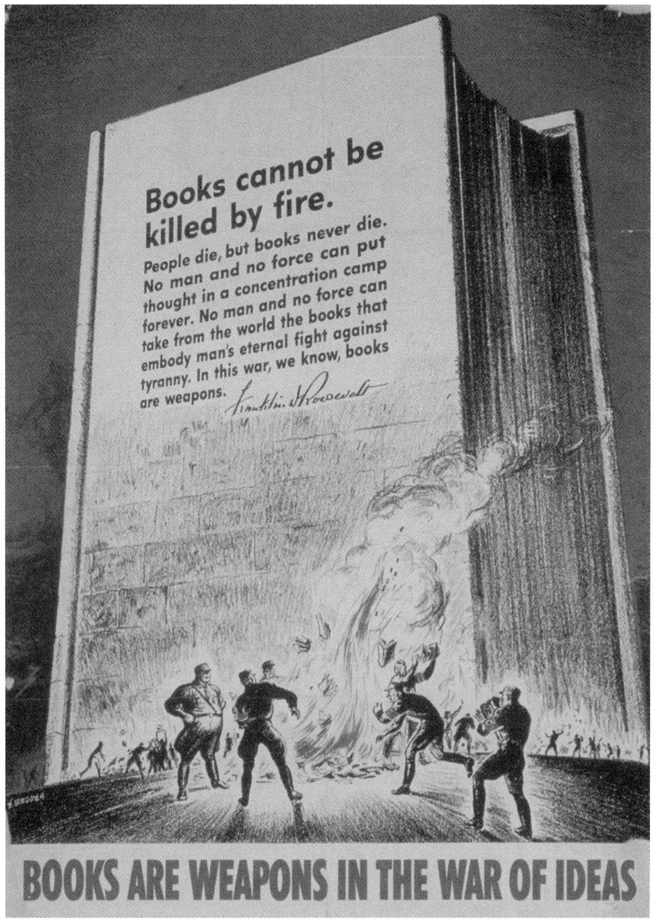

18 US Office of War Information poster, 'Books are Weapons in the War of Ideas' (1942).

then they used it infamously to burn books. Cordoning off the courtyard of Berlin's Humboldt University, the SA stacked books by Jews, communists and other 'degenerate' writers, including Franz Kafka, Bertolt Brecht and Albert Einstein, and set them alight. The colour lithograph (a propagandistic image made years later at the height of the Second World War) by Steve Broder (1906–1992) shows a giant book towering like a massive gravestone over SA men who gleefully throw volumes onto a bonfire (illus. 18).

That Semper's grand synagogue should be destroyed by fire at the hands of the SA is one of those (frightful) ironies of history. Liberally minded, Semper saw fire as one of the four essential elements of architecture: his was an understanding essentially beneficent and utopian:

> Before men thought of erecting tents, fences, or huts, they gathered around the open flame, which kept them warm and dry and where

> they prepared their simple meals. The hearth is the germ, the embryo, of all social institutions . . . It was a moral symbol: it joined men together into families, tribes, and nations . . . The hearth has kept its age-old significance up to the present. In every room the center of family life today is still the fireplace.[45]

Semper is not alone in his fascination with the hearth: much earlier, Leon Battista Alberti's treatise on architecture, *De re aedificatoria* (*On the Art of Building in Ten Books*, 1452), framed fire in ways that drew attention to its intimate and private, yet public and social, dimensions (see chapter Three).

The 'Greatest Fire' in Modern History: Chicago in 1871

> What will doubtless prove the most destructive conflagration of modern times has been raging for two days and nights in our sister city of Chicago, laying waste alike the dwellings of the poor and the palaces of the rich . . . The city, which at the close of the week was bustling with prosperity . . . is now for the greater part a lurid waste. A hundred thousand are homeless . . . It is the GREATEST FIRE IN MODERN HISTORY . . . We do not except even the GREAT FIRE OF LONDON. For the aggregate losses will undoubtedly be greater now than then.
> *St Louis Times* (11 October 1871)[46]

And this was the case: the fire ravaged 18,000 buildings, the homes of nearly 100,000 Chicagoans were destroyed, 300 were dead.[47] Beginning on Sunday, 8 October 1871, in the barn of Patrick and Catherine O'Leary behind their cottage at DeKoven and Jefferson Streets, the fire moved effortlessly through the wooden, labyrinthine buildings of the West Side, then leapt the Chicago River into the South Side, and into the very heart of the city. Fanned by a powerful, dry wind from the southwest, it was unstoppable.

A burgeoning city in 1871, Chicago contained all the ingredients of modernity. There were cast-iron structures such as the Chicago Tribune Building (supposedly resistant to fire, it would fail to match this promise, and was destroyed). There were railroads, industries and machinery. But there were also many old working-class neighbourhoods constructed of wood, and these helped facilitate the fire. By all criteria, however, Chicago on the eve of the cataclysm was the very model of a modern city. And it is this modernity that perhaps is the single most important reason why the disaster, as the historian Carl Smith has written, 'ignited a literary conflagration as extensive in its own way as the fire'.[48] There are personal accounts of the fire that are redolent of the diaries of Pepys and Evelyn;[49] but there are also the many pieces by journalists and professional writers that give (seemingly) first-hand accounts in newspapers and books. Repeatedly evoked in the same breath are the contrasting notions of the 'marvellous' and the 'dreadful', ideas that translate the tricky notions of the Sublime in ways that struck a chord with a modern readership.

Chicago and the Great Conflagration, written by Elias Colbert and Everett Chamberlin and published one year after the fire in 1872, exemplifies this type of commentary. Editorialized by these two writers, and with various first-hand testimonies (some acknowledged by name, others not), the book is rich in evocation and metaphor. And these make clear fire's snake-like turning-in on itself: declaring, on the one hand, its splendour and spectacle, but on the other its terror and deceit:

> The sensation conveyed to the spectator of this unparalleled event, either through the eyes, the ear, or other senses or sympathies can not be adequately described, and any attempt to do it but shows the poverty of language . . . As a spectacle, it [the fire] was beyond doubt the grandest as well as the most appalling ever offered to mortal eyes

> . . . Added to the spectacular elements of the conflagration – the intense and lurid light, the sea of red and black, and the spires and pyramids of flame shooting into the heavens – was its constant and terrible roar, drowning even the voices of the shrieking multitude.[50]

Elsewhere, Colbert and Chamberlin quote a local writer saying,

> The people were mad . . . They stumbled over broken furniture and fell, and were trampled under foot . . . Liquor flowed like water, for the saloons were broken open and despoiled, and men on all sides were to be seen frenzied with drink . . . Women, hollow-eyed and brazen-faced, with foul drapery tied over their heads, their dresses half torn from their skinny bosoms, and their feet thrust into trodden-down slippers, moved here and there, stealing, scolding shrilly, and laughing with one another at some particularly 'splendid' gush of flame or 'beautiful' falling-in of a roof.[51]

In this stirring account, civilian disorder and aesthetics come together in an overwhelmingly apocalyptic yet wondrous scenario. One is left with the overwhelming impression that it is the spectacle of the fire, its ability to thrill and to inspire, that is the very stuff of the reports. Hence the metaphor of 'painter' is invoked some pages later, where we read that: 'Fire was a strong painter and dealt in weird effects, using only black and red, and laying them boldly on.'[52]

In addition to the literary response, there are the illustrations that appeared in the local and international press. Fanciful as some are, they nonetheless are important constructions of the disaster. Two bird's-eye views from exactly the same viewpoint show sail and steamboats on Lake Michigan and flames sweeping over the city. In the more dramatic of the two, some of the vessels are burning, and the flames rise thickly towards a Baroque-like crescendo of clouds. This is an image of disaster, celestial and awe-inspiring (illus. 19). The other illustration is far more sober: wispy smoke blows across the city, and none of the waterborne craft are damaged (illus. 20).

While the people of Chicago are excluded from these panoramas, they are very much present in an engraving based on a sketch by John R. Chapin (1827–1907), *The Rush for Life over Randolph Street Bridge* (illus. 21). One of the 'special artists' for *Harper's Weekly*, his image was published in that magazine some three weeks after the disaster.[53] It is a scene of bedlam: a throng of people swarms across the iron bridge (named after Charles Randolph, Secretary of the Chicago Board of Trade); horse-drawn vehicles are prominent; and men and women carry their special possessions on their backs. Meanwhile, the fire swirls

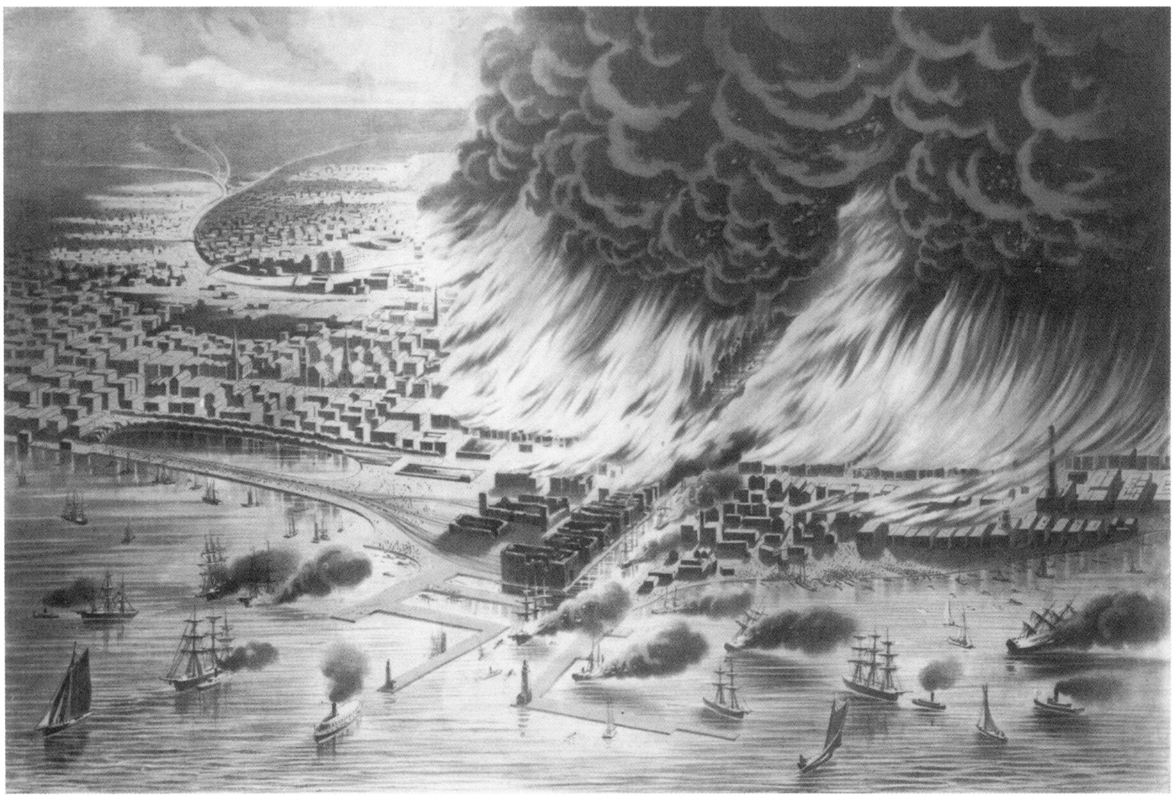

19 William Flint, *Chicago in Flames*, 1871, lithograph.

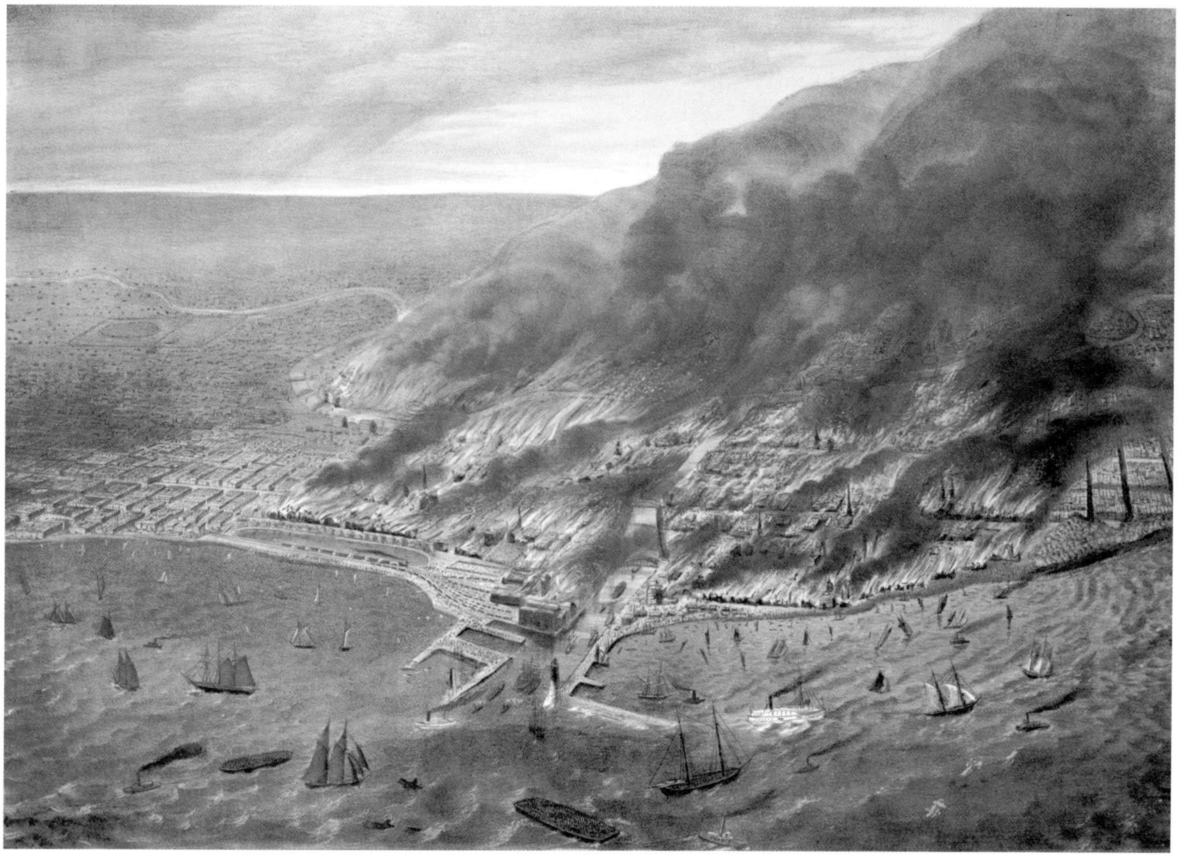

20 Thomas Kelly, *Destruction of Chicago by Fire, Oct. 1871*, 1871, lithograph.

menacingly above them. Chapin would describe these events in a letter published in *Harper's Weekly*:

> I confess that I felt myself a second Nero as I sat down to make the sketch which I send herewith of the burning of Chicago . . . I soon found myself at the Randolph Street Bridge, the point whence my sketch was taken. No language which I can command will serve to convey any idea of the grandeur, the awful sublimity, of the scene. For nearly two miles to the right of me the flames and smoke were rising from the ruins and ashes of dwellings, warehouses, lumber-

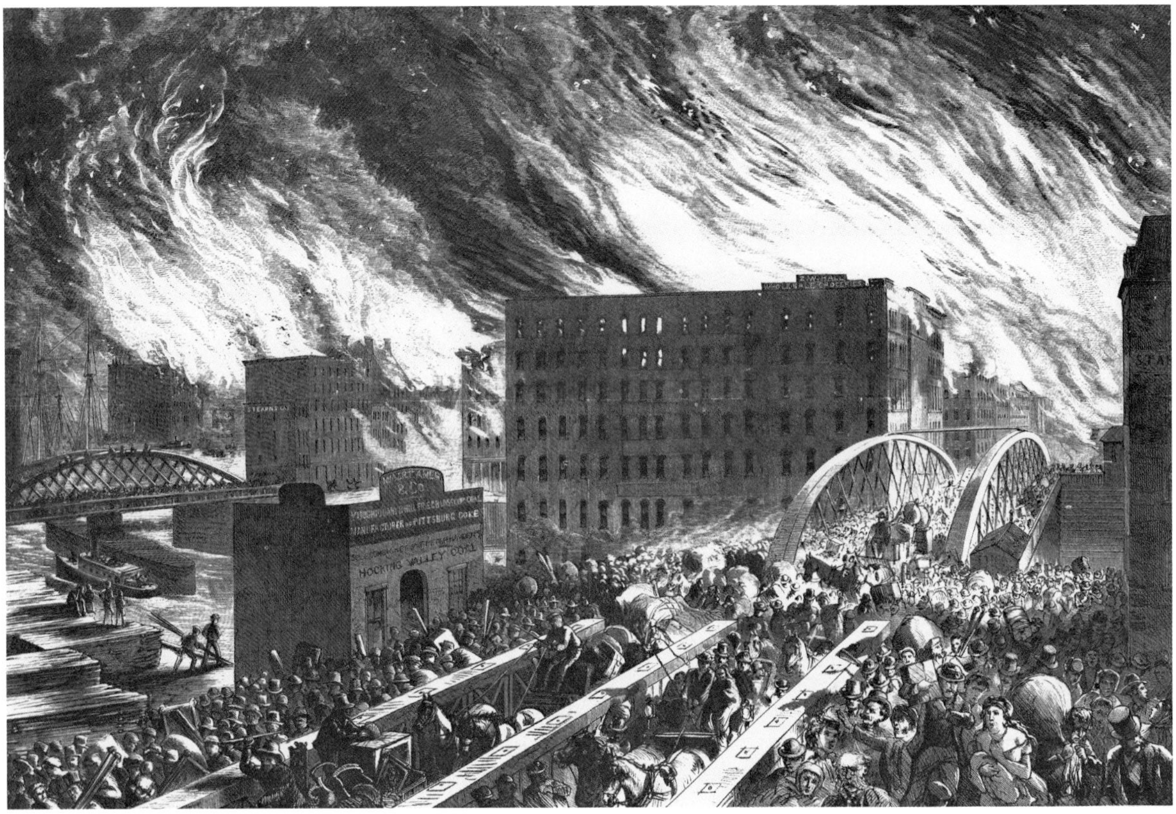

21 Engraving from a sketch by John R. Chapin, *Chicago in Flames – The Rush for Life over Randolph Street Bridge*, 1871.

> yards, the immense gas-works; and the view in that direction was bounded by an elevator towering one hundred and fifty feet in the air, which had withstood the fire of the night before, but which was now a living coal, sending upward a sheet of flame and smoke a thousand feet high.[54]

Chapin's gaze falls on the blaze and its effects on buildings. By contrast, Joseph Edgar Chamberlin, a young journalist who followed closely the development of the fire over the two days of its duration (8–9 October), emphasized the commotion in the vicinity of Randolph Street:

> Wagons were rushing through the streets, laden with stocks of goods, books, valuable paper . . . scores of men were dragging trunks frantically along the sidewalks, knocking down women and children.[55]

(This account, as it so happens, complements Chapin's drawing well.) Chamberlin noted particularly the reaction of onlookers:

> But as large as was the number of people who were flying from the fire, the number of passive spectators was still larger. Their eyes were all diverted from the scurrying mass of people around them to the spectacle of appalling grandeur before them.[56]

An illustration showing the burning of the Crosby Opera House captures this sense of awe and horror. Great wavering, fabric-like lengths of flame emerge from the facade while people rush hither and thither rescuing what they can from the building. In the foreground top-hatted men look on. In a passage that could sit easily with this incident Chamberlin writes:

> They [the spectators] stood transfixed, with a mingled feeling of horror and admiration, and while they often exclaimed at the beauty of the scene, they all devoutly prayed that they might never see such another.[57]

These conflicting emotions in the face of fire – wonder and alarm – bring the Sublime into the realm of the everyday. Something very much like this had occurred in Edinburgh as well, some four-and-a-half decades earlier.

'The Wind was Extremely Gentle': Edinburgh and its Great Fire of 1824

Edinburgh's 'Great Fire' does not occupy the high ground of recognition and commentary of, say, a Chicago or a Dresden. But fire, of course, does not know hierarchies: it is anywhere and everywhere; and everywhere it has the potential to create legends.

And so it did in Edinburgh. Beginning on the evening of Monday 15 November 1824, shortly before 10 pm, in a house on Old Assembly Close, High Street, the fire soon attracted 'several fire-engines' and even 'The Sheriff, Lord Provost, and Bailies'.[58] Unlike London in 1666, and Chicago in 1871, Edinburgh did not have winds to fan the flames. 'The wind', on the contrary, 'was extremely gentle, and came from the south-west'.[59] Still, the fire spread from the tenements of Old Assembly Close to the Courant Office. Robert Chambers describes these events in ways that part company with the melodramatic accounts of the burning of London and Chicago:

> The night was calm and serene, and the sparks sent forth by the flames rose high into the air, like embers from the crater of a volcano. Their appearance was like the thickest drift of a snow-storm – an image that was suggested to the fancy of many of the by-standers; and the writer of this narrative was particularly amused by hearing a native of Aberdeenshire exclaim, 'See the red snaw!'[60]

This likening of fire to snow, and the striking reversal of the adjectives 'white' and 'red', are imaginative constructs that take fire into the realm of poetics. As the evening wore on 'the wind, accompanied with rain, rose in sudden and fitful gusts'.[61] By early Tuesday morning, however, 'the fire had considerably abated, and by mid-day it seemed to be nearly subdued'.[62] This was the lull before the storm. Shortly after noon that day, the Tron Church (built in 1641) in a neighbouring district caught alight. Chambers writes about this blaze in ways that recall some of the rhetoric surrounding London's Great Fire. On that occasion, the date 1666 was linked to 666, the 'Number of the Beast'; and the fire was associated with Revelation's prophecies of the Second Coming.[63] Similarly, some in Edinburgh saw the burning of the Tron Church – 'something sacred' – as 'judgement-like!'[64]

For Chambers, the destruction of the church was 'a spectacle singularly terrific and sublime'.[65] Again, that word 'Sublime'. As the fire spread over the next few days to Parliament Square, St Giles and beyond, it would be described in similar terms. As the *Edinburgh Evening Courant* evidences:

> The scene was now awfully grand; and could we have divested ourselves of the thoughts of the losses, and hardships and ruin, which attended the progress of the conflagration, we could not have been placed in a situation where we could have derived such a portion of sublime enjoyment.[66]

Likewise, *The Scotsman* reported on the third day of the fire, a Wednesday:

> We have described the effects of the fire as they were seen from our own office. From a greater distance we are assured the effects were awfully grand. Abstracted from the sense of danger to friends and neighbours, the broken masses of the old town, its chimneys and spires, now thrown into shade, now fiercely illuminated by the raging flames, and seen through showers of living fire, presented a scene at once wild, terrific and sublime. A nearer approach changed the feeling. Pity was excited by the condition of the poor – men, women and children, driven from their homes.[67]

Edinburgh's Great Fire lasted for four days. Thirteen were killed (this number is still contested) and some 400 homes, and a large area of the city, were destroyed. Compared with some of the other fires discussed in this chapter, these statistics are slight. But the affects of fire know no degree. This one was no exception; it entered into the national imagination, and became known as 'The Great Fire of Edinburgh'.

A postscript: James Braidwood was the 23-year-old (and newly appointed) Superintendent of Fire Services when the blaze began in 1824. He would go on to become the first Head of the London Fire Engine Establishment nine years later. Braidwood's Record of Fires in Edinburgh from 23 October 1824 – 19 March 1829, commenced soon after his appointment, has (only) twenty-one hand-written lines on the fire. He describes the blaze in the Tron Church: 'The whole of the wooden part of the steeple was consumed and the roof of the church considerably damaged by the fall of the burning timbers.' He continues with the fire in Parliament Square and then the one 'in some wooden lands between the Old Assembly Close & Carrs Close, but [they] did very little damage'. To conclude this short report he states that 'there were two firemen killed and eleven hurt – beside other people'.[68]

Written, one assumes, in the immediate aftermath of the fire, Braidwood's underestimation of lives lost is understandable. More surprising, however, is his omission of dates; Arthur Pordage, Edinburgh Firemaster from 1896 to 1927, would fill these in 120 years later.[69] His intervention appears in the report's margin (signed and dated by him: '7/2/44') as well as in the body of the text. Pordage is unreliable, however. He gives the date of the fire in Parliament Square as 15 November; it was in fact the 16th. He inserts the dates 14 into Braidwood's 'from Monday the . . . Nov. 10 pm', and 16 into 'till Wednesday the . . . Nov. 12 pm'. Both are incorrect. All this, to be sure, reminds us of the instability of history's chronicling. But it is Braidwood's (presumably) on-the-spot descriptions that, in the end, are the more telling. Prosaic, to say the least, they give a picture of the fire that contrasts strongly with both contemporary and subsequent accounts. As always, reality, rhetoric and myth co-mingle in experience and remembering: The GREAT FIRE is always greater than itself.

This chapter might have ended here if it were not for a poem that, like fire itself, flared up from nowhere (I came across it by chance in a daily newspaper, *The Australian*).[70] Composed by the writer and critic Clive James, well known in the country of his birth, Australia, as well as in the UK where he resides, it is called 'Fires Burning, Fires Burning'. Combining suffering and sorrow in prose that is simple but terrifying, the poem's 182 words belie its vast sweep: from the dying years of the Second World War, to New York and 9/11 and, finally, to the cremation of James's mother in Sydney in 2003, where a tone of solace is introduced into what otherwise is a short, modernist history of the horrors of fire brought about by human agency. To take one instance: the US bombing of Tokyo on 9–10 March 1945, which James refers to in stanza three, where approximately half-a-million M-69 incendiary devices were dropped, destroying some 26 square km (16 square miles) of the city. Coming four weeks after the Allied bombardment of Dresden, the attack resulted in the death of close to 100,000 people, overwhelmingly civilians, and constituted probably the most dreadful air raid in history.[71] Remarkably, these bombings led to a reputedly greater immediate loss of lives than the atomic attacks on Hiroshima and Nagasaki.[72] Where else to insert Clive James's poem, then, but here:

> Over Hamburg in March 1943
> The Lancaster crews could feel the heat
> Through the sides of the aircraft.
> The fire was six thousand feet high.
> Further East, there were other fires.
> When burning a lot of bodies, the SS found
> The thing to do was to put down a layer
> Of women first.
> They had more fat in them.

In Tokyo, on the night of March 10th 1945,
Some people who survived in a canal
Saw a horse on fire running through the streets.
But few who saw it were left to remember anything:
Even the water burned.

In New York, on September 11th 2001,
Some couples, given the choice
Between the flames and a long fall,
Outflanked the heat and went down holding hands.
Come with me, you imagine the men saying,
I know a quicker way.

Sydney, two years later. Next to my mother's coffin
I gave thanks that she would shortly meet
A different kind of fire,
Having died first, and in due time.

two

Dreaming Fire/Shedding Light

To burn it ['The Temple'], it's like giving it a higher force . . .
It's like an offering.
Fred Dickson, assistant in the building of 'The Temple',
Burning Man Festival, US, 2004.[1]

The Great Fires discussed in the previous chapter destroyed structures and lives. They also went beyond this into that more indeterminate space of rhetoric, reconstruction and memory. At times, the minutiae of these fires were singled out for special comment. At others, fire's overwhelming impression was commented upon, and nearly always in terms of a heady mix of incomprehension, horror and awe. Not so with Dresden, however, where that all-mighty firestorm and its horrific results demanded a very different response. Take this of the firefighter Alfred Birke, pithy and poignant:

> Flames shoot from the facades of buildings . . . I don't meet a living soul. At the Pirnaischer Platz I encounter three naked bodies, a woman and two children. I take care not to drive over them. At last the smoke thins a little, the flames retreat.[2]

In all of this, fire's ancestry in the complex and conflicting lexicon of myth and religion was largely overlooked. Understandably, since these blazes and their impact on citizens (during times both of war and peace) demanded a response and an explanation that suited the moment. But fire is never that straightforward. It may wreak havoc on the 'man-made', that much is clear, but its effects on bodies is something else: disfiguring or reducing them to ashes, it also has the singular ability to warm and to comfort them. And in each and all of these outcomes, fire illuminates. It sheds light, literally. That this mercurial element occupies a central place in many belief systems and their representations should come as no surprise. That it would allow Buddhist monks to self-immolate during the Vietnam War or, in striking contrast, allow participants in Burning Man, an annual festival in the Nevada Desert, USA, to celebrate the (supposed) beneficent powers of fire, simply reinforces fire's enduring capacity for paradox and permit.

Aboriginal Dreamings

Understandings of fire and its relationship to the real and the imaginary life commence with the indigenous peoples of Australia, some 50,000 years BC, where it is inextricably linked, like all other natural phenomena, to Dreaming. Aborigines use this European term to describe 'the spiritual, natural and moral order of the cosmos', as Wally Caruana, a specialist in Aboriginal art and culture, has put it.[3] The term has no associations whatever with dreaming in general or the realm of the unreal. To quote Caruana again, the Dreaming 'relates to the period from the genesis of the universe to a time beyond living memory . . . [and] focuses on the activities and epic deeds of the supernatural beings and creator ancestors . . . who, in both human and non-human form, traveled across the unshaped world, creating everything in it and laying down the laws of social and religious behaviour.'[4]

Aboriginal Dreamings of fire variously combine the cursed and the blessed, the deceitful and the faithful – and the destructive and redemptive; they do so in ways that anticipate all later symbolic configurations of fire, notably the myth of Prometheus. One of their original fire myths involves a cold and inhospitable earth where a man and his wife see a glow brought about by a lightning strike on a piece of wood. Nonplussed, they attempt to conceal this strange thing but it suddenly breaks out into flame. Calling upon an elder to destroy the fire, they return with him to the scene and, unexpectedly, find it comforting; this alerts the elder to the beneficent properties of fire, and he proceeds to distribute amongst the community its twofold elements of light and warmth.[5]

Dreamings cover the length and breadth of Australia. Some are distinct to specific places and belong to those who live there, while others spread across huge distances and link the inhabitants of the lands they travel over. Warlukurlangu, meaning 'belonging to fire' in Warlpiri, is named after a Fire Dreaming associated with a stretch of land near Yuendumu in the vicinity of Alice Springs, Central Australia. This Dreaming is 'managed' by Dolly Granites Nampijinpa Daniels (also known as Dolly Daniels Granites Nampijinpa, *b. c.* 1935), whose language group is Warlpiri, and who lived and died in Yuenmendu in 2004. Warlukurlangu brings together fire with the themes of deceit, punishment and redemption.

As the story goes, a cunning old man named Jampijinpa Bluetongue fooled his two sons into believing he was blind. As a consequence, they would hunt on his behalf while he stayed at home indulging his pet kangaroo. A recent children's illustrated book, *The Magic Fire of Warlukurlangu*, continues the tale:

> Because they thought he could not see, the two sons looked after their father very well . . . After a while, [they] became tired of hunting every day for their father. One day, without his knowing it, the two boys speared their father's pet kangaroo and cooked it on their fire. Then they . . . served it up to him.[6]

Delighted by its taste, the father had 'another piece and he ate it up'. But after a few days, he began to worry about the absence of his kangaroo.

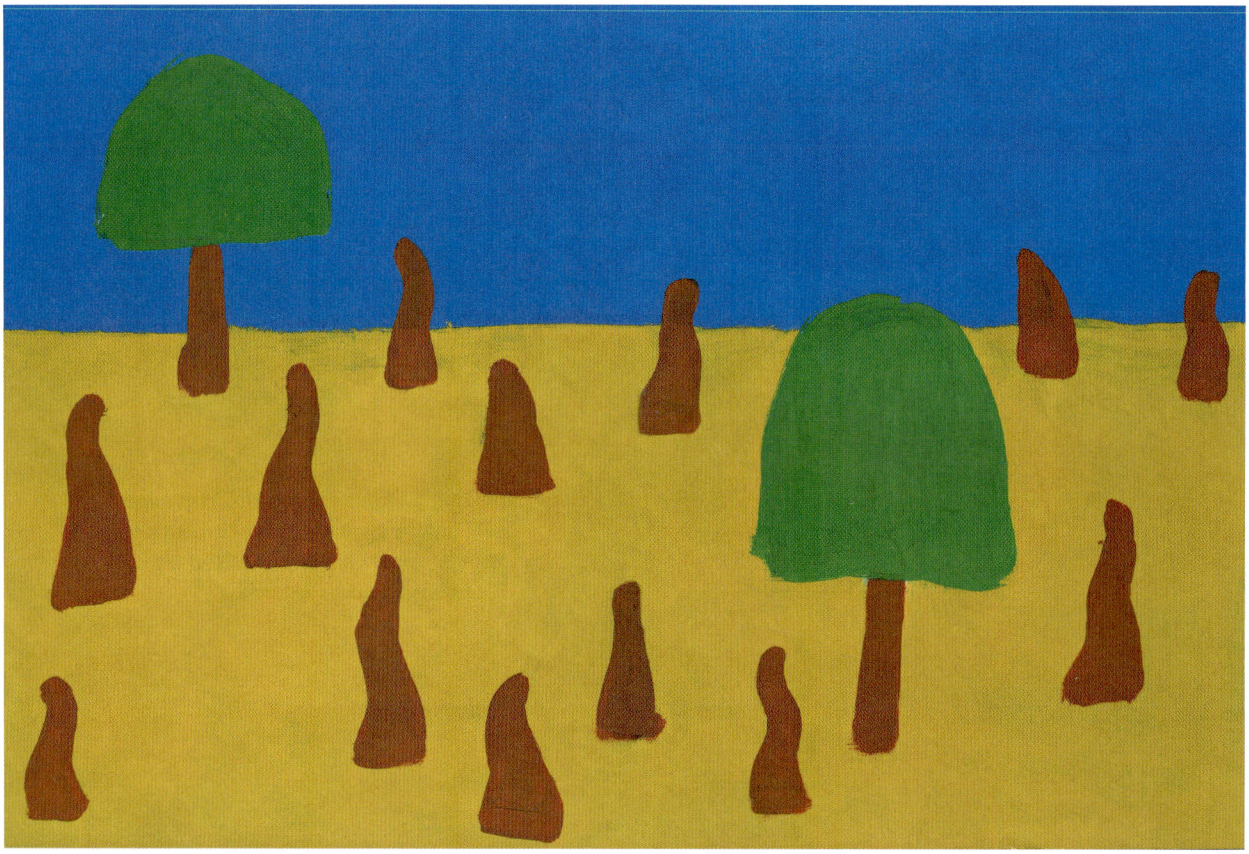

22 From Dolly Daniels Granites Nampijinpa, *The magic fire at Warlukurlangu: a Dreaming narrative / belonging to Dolly Daniels Granites Nampijinpa*, series edited by Christine Nicholls assisted by Sue Williams (Kingswood, South Australia, 2003).

Finally realizing what his sons had done, 'Jampijinpa made a magic fire that was awaiting for them when they returned home'. At first, thinking their father was in danger, the sons tried to put it out. 'But every time they put it out, the magic fire would light itself again.' They tried to escape the fire, but were killed by its 'smoke and flames'.

Today the flames from this fire Dreaming can still be seen in the Warlukurlangu region in large anthills (illus. 22), which, it is believed, resemble the shape of flames, and serve as a reminder to the Warlpiri people of what happened there.

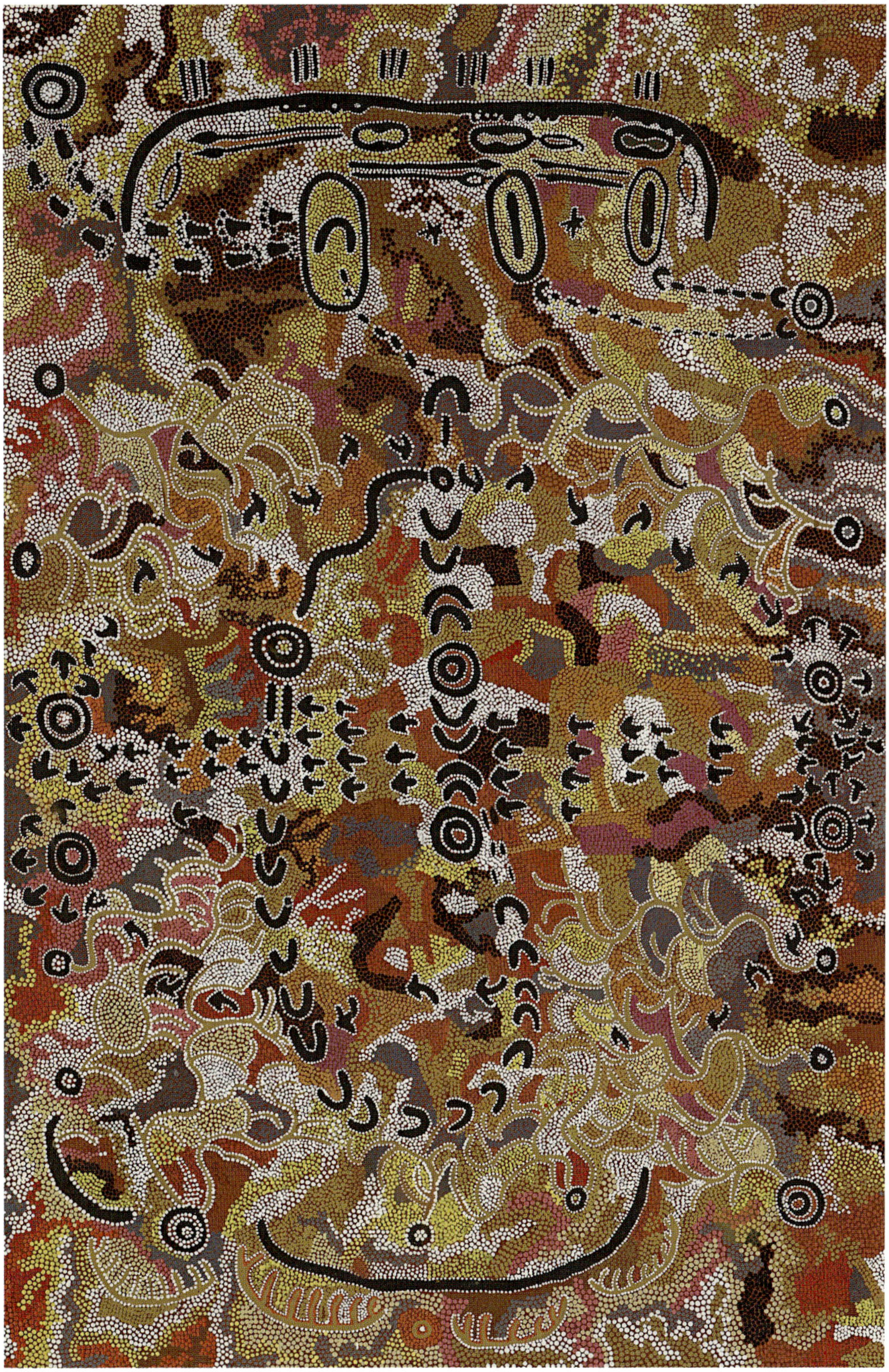

23 Uni Nampijinpa Martin (Warlpiri, born *c.* 1942), and Dolly Nampijinpa Daniels (Warlpiri, *c.* 1935–2004), *Warlukurlangu Jukurrpa (Fire Country Dreaming)*, 1988, synthetic polymer on canvas.

This Dreaming is about a double deception: enacted by a father upon his sons, and they on him in return – and its fateful results. In a painting made in 1988, now held in the National Gallery of Victoria, Australia, this Dreaming is translated into a complex network of signs and patterns by the same 'owner' of the story, Dolly Nampijinpa Daniels and Uni Nampijinpa Martin (*b. c.* 1942), also a Warlpiri woman (illus. 23). One example, among many others, of contemporary Aboriginal painting that now is secured in the *lingua franca* of the artworld, this work illustrates the changed and changing circumstances of indigenous peoples and their cultural productions. It is a contested history, to be sure, but one that lies beyond the scope of this study. Suffice it to say that it includes the pioneering work produced in the 1970s by artists of the Papunya Tula Painting Movement of the Western Desert in Northern Australia, under the auspices of a European schoolteacher, Geoffrey Bardon, who would go on to write three seminal books on the subject.[7] For our purposes, this defining moment in indigenous art provides telling examples of fire and its nuanced relationship to Aboriginal life: to cite just two – paintings by David Corby Tjapaltjarri (*c.* 1940–1980) and Johnny Warangkula Tjupurrula (*c.* 1925–2001). A Warlpiri man, Tjapaltjarri's *Woman's Fire Dreaming* (illus. 24) shows women cooking goannas or kangaroo ('bush tucker'). Concentric ovals in the painting establish the large fireplace while 'the radiating lines are [smouldering] firesticks that the women will whirl above their heads in the ceremony'.[8] (Used commonly by Aboriginals in their travels to ward off flies and to help heat the body, 'fire-sticks' would come to designate an ingenious type of Aboriginal fire management known as 'fire-stick farming', discussed below).

Woman's Fire Dreaming is a celebration of the simple, everyday demands of cooking and nourishment. So too is the watercolour by the English convict artist, Joseph Lycett (*c.* 1775–1828), which in very different ways, of course, paints an idyllic scene of Aboriginals fishing and cooking by moonlight: those on the boats carry burning sticks

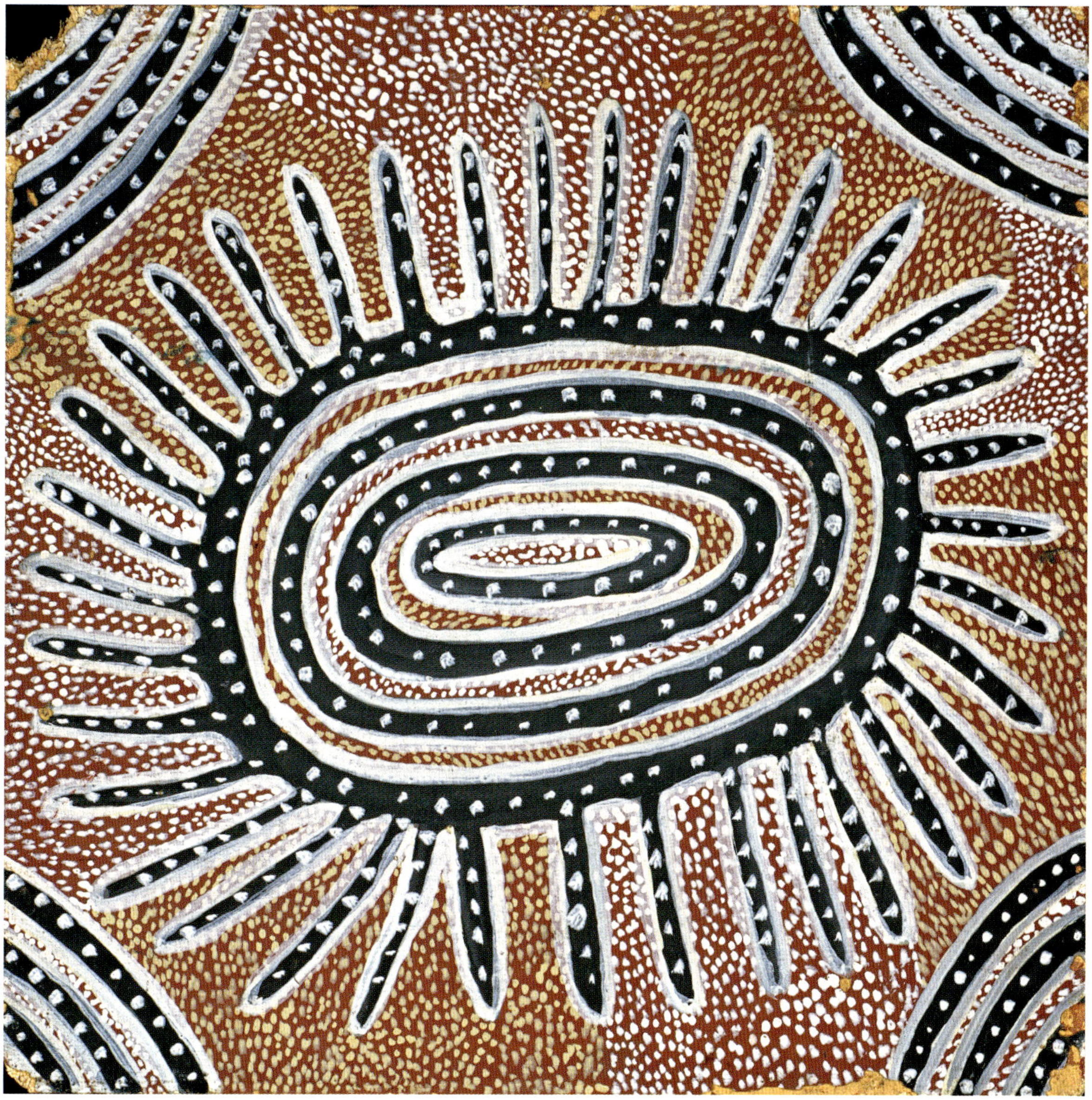

24 David Corby Tjapaltjarri (Warlpiri, *c.* 1940–80), *Woman's Fire Dreaming*, 1973, poster paint with PVA Bondcrete glue on hardboard.

25 Joseph Lycett, 'Fishing by torchlight, other Aborigines beside camp fires cooking fish', *c.* 1817, watercolour from *Drawings of the Natives and Scenery of Van Diemens Land 1830*.

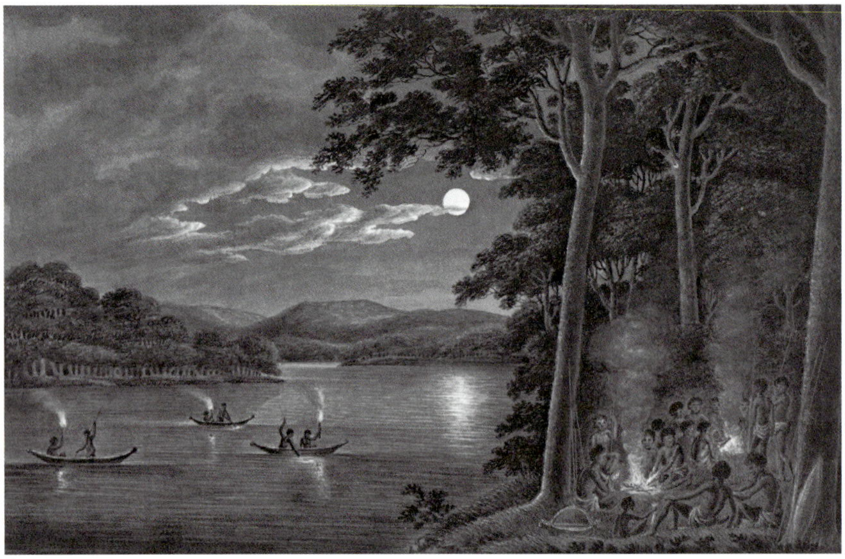

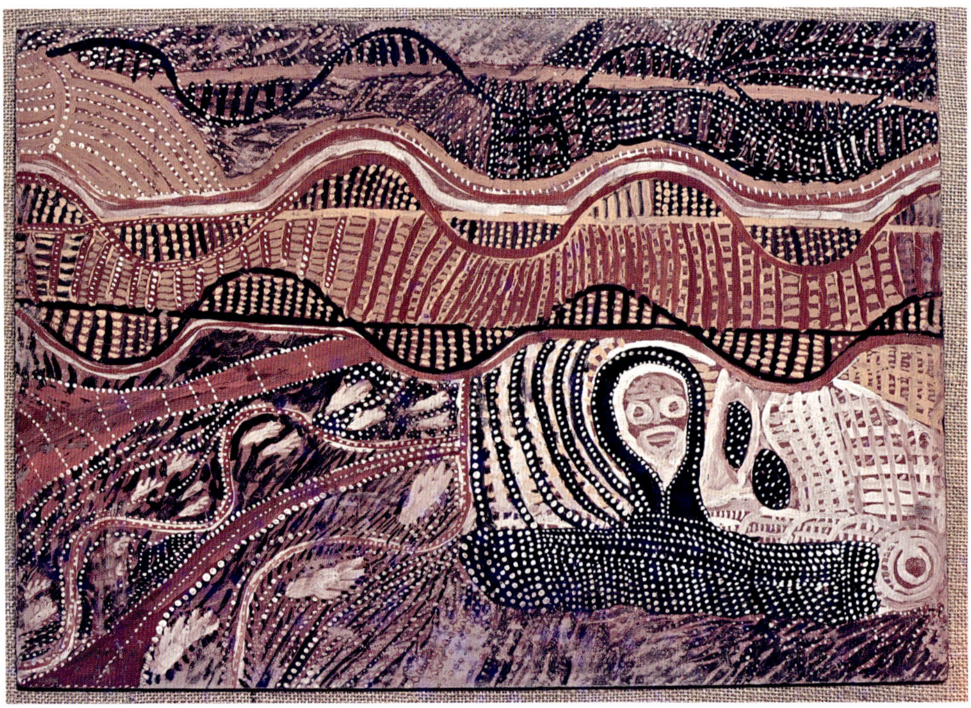

26 Johnny Warangkula Tjupurrula (Loritja, *c.* 1925–2001), *Bungalung Old Man's Fire Story*, 1972, poster paint with PVA Bondcrete glue on canvas.

while a mixed group of women, children and men cook fish by fire on the bank (illus. 25). A more sinister theme, however, unfolds in Johnny Warrangkula Tjupurrula's *Bungalung Old Man's Fire Story* (illus. 26).[9] A member of the Loritja tribe, the story Tjupurrula relates concerns an old man called Bungalung, a giant, chased by wallaby (mala) spirit people. Before they reach him, however, he is caught in a fire. His presence in the painting is denoted by the (stylized) wide-eyed face towards the bottom left; the black patterning at the top represents the burnt bush; and the cream-coloured tracks, some large and some small, indicate Bungalung's attempts to flee the encroaching flames. When the wallaby spirit people eventually find him, he is still alive. Later he dies.

A work that contrasts strongly with these two images is *Bushfire* (2003), by the Pintupi painter Ronnie Tjampitjinpa (*b. c.* 1943, illus. 27). By the early 1980s, writes Wally Caruana, the Pintupi

> had begun to return to their ancestral lands around Kintore and Kiwirrkura in the west [and] began to develop archetypal images constructed purely of concentric circles representing sites with journey lines joining them.[10]

Making 'minimal use of iconography to imply a wealth of concepts far deeper than the interpretation given to the uninitiated', their works are distinguished by an austere formalism.[11] This is evident in Tjampitjinpa's *Bushfire*, where circles give way to five black lines resembling incisions that traverse, in part, a painstakingly painted surface made up of ochre colours. Denying Western perspectival space, *Bushfire* for the 'uninitiated' (to use Caruana's term again) suggests the awful symmetry of fire seen from above.

In its strictly pared-down forms *Bushfire* may be compared to two paintings respectively by an Australian colonial artist, the Austrian-born Eugene von Guérard (1829–1890) and a contemporary white

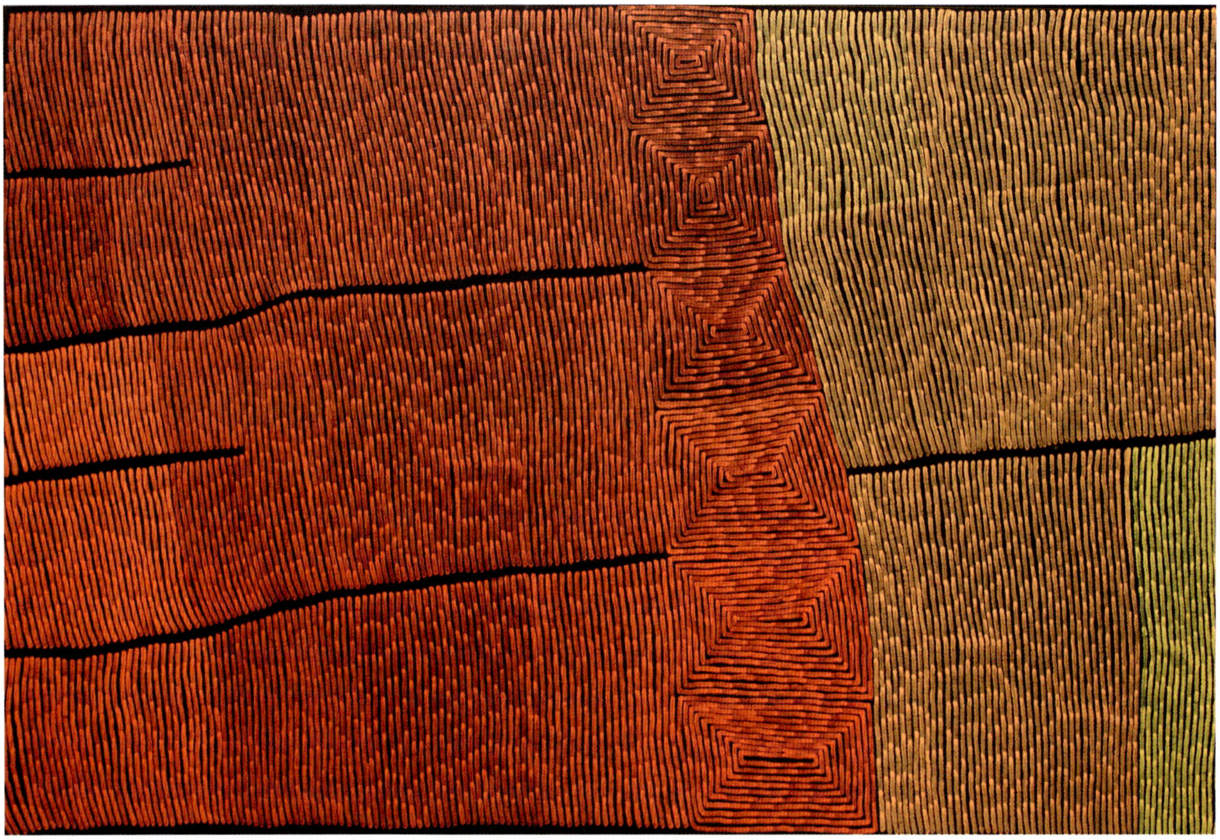

27 Ronnie Tjampitjinpa, *Tingari Bushfire Dreaming*, 2003, acrylic on canvas.

Australian artist, Shane Fitzgerald (b. 1973). Like Tjampitjinpa's, both their works negotiate the horizontal of the canvas in ways that emphasize stretches of fire. Von Guérard's *Bushfire between Mount Elephant and Timboon* (illus. 28) has as its backdrop a moonlit scene; in the foreground a fire sweeps inexorably from left to right across the length of the horizon, sending columns of smoke into the sky, and separating heaven and earth, as it were, in one neat line. Devastation is implied – of humans, animals and nature – but what we are left with is its opposite: an awe-inspiring, fiery magnificence, in which the heavily foliated trees silhouetted at the extreme left stand defiant. Here we are back in the world of the Sublime, so powerfully evoked in the images and the rhetoric of the 'fabled fires' discussed in chapter One. Shane Fitzgerald's large photograph *Inferno, Arnhem Land*, meanwhile, shows what seems to be molten lava stretching across the length of the canvas's rectangular format (illus. 29). Unlike the von Guérard, this hot-to-touch stream of fire is located at the bottom reaches of the picture, and wispy veils of smoky reds and blacks spread throughout the remainder of the image.

'Fire-stick farming', referred to above, was an Aboriginal response to the extraordinarily flammable nature of the country. Involving the lighting of fires at critical times to help replenish the bush and attract animals to re-sprouting grasses – a striking instance of fire's redemptive powers, literally – this policy, as Gavin Wilson has recently observed,

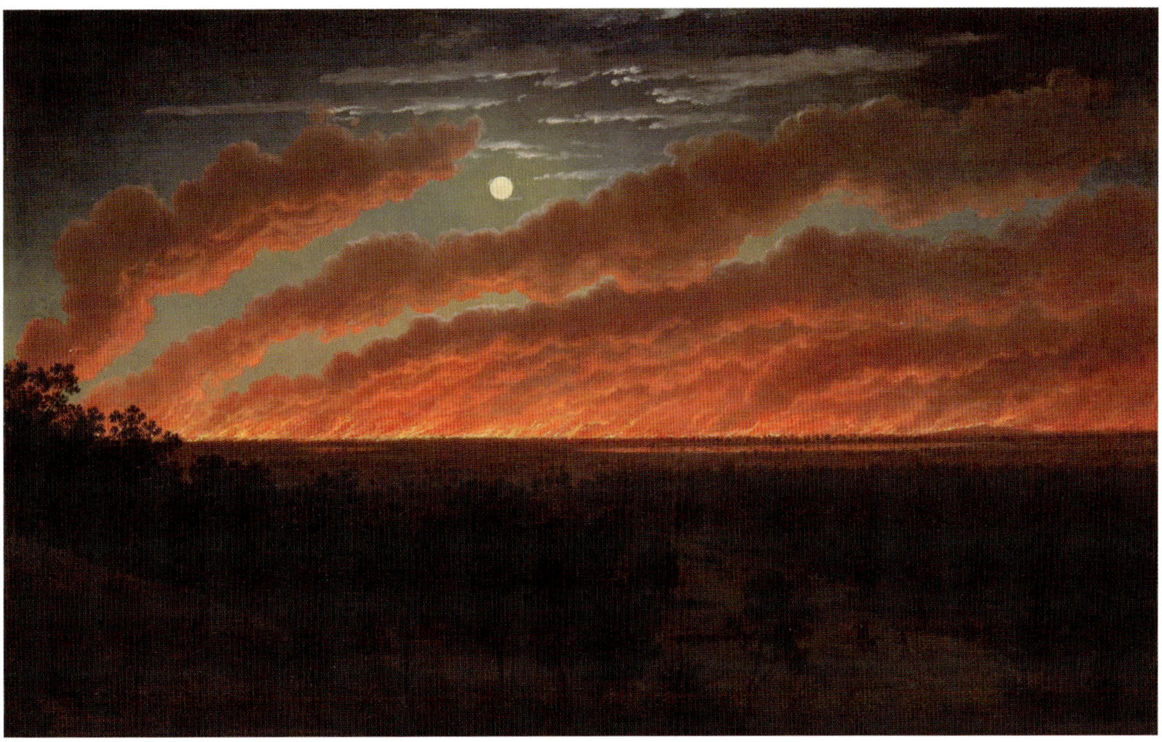

28 Eugene von Guérard, *Bushfire between Mount Elephant and Timboon, March 1857*, 1859, oil on canvas mounted on board.

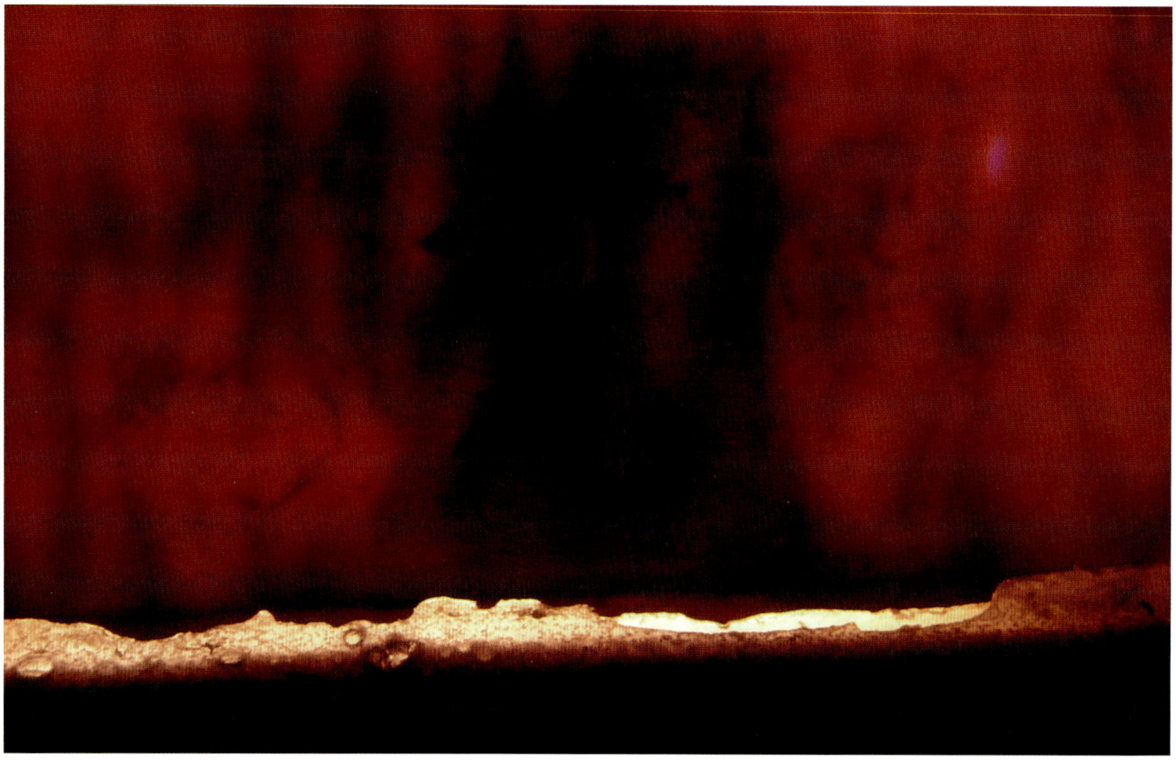

29 Shane Fitzgerald, *Inferno, Arnhem Land*, 2002, duralux print.

would be the critical factor in the release of nutrients into Australia's woodland and grassland ecosystems. The widespread practice maintained a viable food supply for the Aboriginal population. As well, the fire-loving species of plants, such as banksia and hakea, that evolved in the vast hinterlands, relied specifically on fire for the release of seed and pollen.[12]

Fire, in short, was at the very heart of Aboriginal societies. In ceremonies, it served both a practical and symbolic purpose. Taking place largely at night, writes Margaret Clunies-Ross, these ceremonies relied on fire to illuminate them.[13] She draws attention to early European artists such as Charles-Alexandre Lesueur (1778–1846) who provided one of the earliest depictions of a ceremonial performance, a 'Corroboree', in which fire plays a central role. Silhouetting a group of seated figures in the foreground and a large tree-trunk, its flames rise and smoke wafts over the gathering. Fire, however, was something that could also be manipulated. Says Clunies-Ross again:

> It may be made to flare up suddenly, by throwing grass or leaves on it; or it may be damped down to smoulder and flicker, so that objects, actors or actions are only dimly visible or visible to some participants but not others. This latter kind of effect is often produced in secret rituals in which it is important that some members of the community hear but do not see what is going on.[14]

These interventions in fire invite comparison with some of the much later activities of performance artists in the 1960s and '70s, discussed in chapter Four, who turned to fire and its fertile associations. But their circumstances and histories, needless to say, would be of an entirely different order.

Recognizing the centrality of fire amongst Aboriginal peoples, the Sydney Paralympic Organizing Committee (SPOC), chose

> a unique way of lighting its Paralympic flame . . . Unlike the Olympic flame and Olympia in Greece, the Paralympic flame has no 'ancestral home' allowing each host city to choose a lighting site of national significance. The flame of the 2000 Paralympic Torch Relay will be lit on October 5 during a traditional fire lighting ceremony on the forecourt of New Parliament House and will be conducted by members of the Ngunnawal community, who are the original inhabitants of the Canberra area [the site of Australia's Parliament]. The Paralympic flame is the essential component of the Torch Relay just as fire was a vital tool in the everyday life of Aboriginal communities before white settlement. Dr John Grant, President of SPOC

> said: 'Igniting the flame by the ancient techniques of the original inhabitants of this land [by friction, with fire sticks] will remind us that while the Paralympic movement is itself young, many of its tenets have much in common with the ancient ideals of the Olympic Games.'[15]

Fire is always a conflicted ground. And so it was at the 2000 Sydney Paralympics (18 September – 9 October) when good-willed politics embraced fire in ways that recognized its essential role in indigenous life and also its relevance to the history of the Olympic Torch Relay. That history goes back many years, but not to the original Greek Olympics, where there was no relay. There were, however, heralds, who were sent out to make sure that warring Hellenic states maintained a truce for the duration of the Games. And it is this that would become the inspiration for the Torch Relay, the first of which was held at the Berlin Games in 1936.[16] Sixty-four years later, in the Sydney Summer Games of 2000 (15 September – 10 October) the flame would visit many Pacific islands, including Guam, the Solomons, Samoa, Fiji and New Zealand, before reaching Australia, where a diver with a specially designed torch transported it underwater at the Great Barrier Reef. Four years later, at the 2004 Athens Summer Games, the torch went on its longest journey ever, travelling some 78,000 km (48,000 miles) throughout all five continents, including two extended sections in Greece itself, where it commenced at Olympia and ended at Athens.[17]

Ancient Greece and the Promethean Condition

Fire pervaded all aspects of ancient Greece where its incarnations and symbolic meanings were multiplicitous, overlapping and typically contradictory. Present at many sacrificial sites, it was embodied in such deities as Hestia, the goddess of the hearth, home and family,[18] Apollo and Helios. The son of Zeus and Leto, and the twin brother of Artemis, Apollo was the god of the sun, logic and reason. In this configuration, he brought together opposing impulses: on the one hand, the fiery nature of the sun, inexplicable and majestic; on the other, the very human attributes of purpose and explanation. Helios, meanwhile, was best known as the sun god. Residing high in the sky, he came to represent the all-seeing witness.[19] Here Helios and, by implication, fire itself, becomes implicated in what we may now describe as surveillance. Fire's 'shedding of light' is never as innocent as it seems.

And so it was with that most mercurial of creations, Prometheus, one of the Titans, and probably the best-known embodiment of fire's mythic properties. Gaston Bachelard (1884–1962) has observed in an (incomplete) book, *Fragments of a Poetics of Fire* (1988, trans. into English 1990), that:

> In his various guises, Prometheus has become a commonplace of human sensibility, universally understood and yet, despite his great renown, essentially unknown. The figure of Prometheus, though dominant and emblematic, seems a paradox, its power fractured into many images.[20]

It is Aeschylus (525/4–456 BC), the earliest of the three famous Athenian tragedians, to whom we owe the literary construction of Prometheus. Of his trilogy about the god – *Prometheus Bound, Prometheus Unbound* and *Prometheus the Fire Bringer* – the first is the only extant text, and even then some have expressed doubts about its authenticity. In *Prometheus Bound*, Aeschylus brings together a motley group of characters: Kraptos and Bia (personifications of Strength and Force), Hephaistus (Zeus' goldsmith; known to the Romans as Vulcan), Oceanus (the eldest Titan, and the personification of the river that ran around the known world), Io (or Lo), a priestess of Hera, much loved by Zeus and transformed into a cow, and Hermes (Mercury, god of heralds, travellers and thieves). A chorus of ocean nymphs concludes Aeschylus' script which, in short, has Prometheus creating the first man from clay and water and then, to make his life more tolerable, stealing fire for him from heaven. These audacious acts had their price: the all-powerful god, Jupiter (known to the ancient Greeks as Zeus), realized that his temporal powers would be compromised by the advent of humans. He would later initiate a counter-offensive, most notable of which was his letting loose of the beautiful Pandora whose female charms, when unleashed from her 'box', wreaked havoc.[21] Poor Prometheus, meanwhile, was shackled to a rock in the Caucasus where, each day, an eagle bit at his liver; it was restored again every night. (There are echoes here of the Hindu god of fire, Agni, discussed below, who is considered to be both cruel and compassionate. Like Prometheus, he has the ability to be reborn: Agni is reincarnated daily from the machinations of holy maidens who twirl the sacred fire drill, a means of creating fire by friction.[22])

In these bare bones of the myth, Prometheus is constructed in ways that collapse inventor, egalitarian, thief and victim. It's a nuanced story filled with twists and turns and contradictions. As might be expected, it has inspired many artists throughout the centuries. The Utrecht-born Dirck van Baburen (*c.* 1594/5–1624), a pupil of Paulus Moreelse, and one of the group of artists known to us as the Utrecht Caravaggisti, shows a muscular (and life-sized) Prometheus in *Prometheus being Chained by Vulcan* (illus. 30).[23] Foreshortened, he lies spread-eagled on the floor of a forge, his creased brow and wide-open eyes registering his agony. The eagle, which is to pick at his body, hovers menacingly at the top left of the painting. Holding a chain in one hand, the blacksmith god

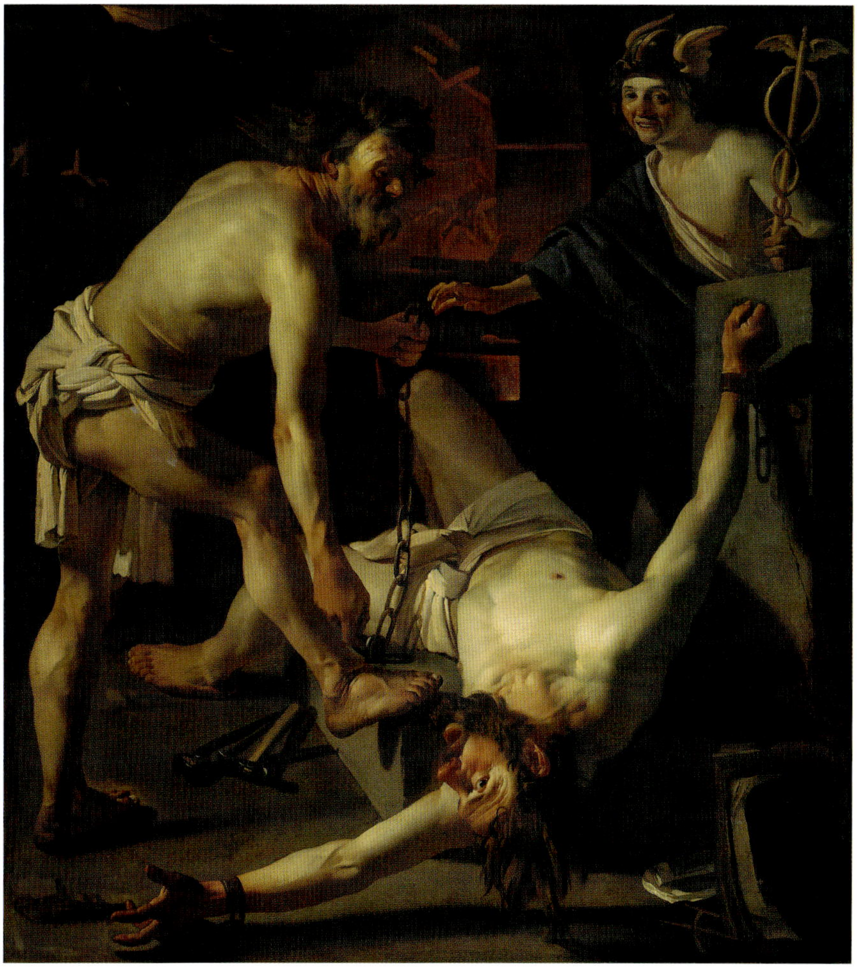

30 Dirck van Baburen, *Prometheus being Chained by Vulcan*, 1623, oil on canvas.

of fire looks down disinterestedly at Prometheus: an interrogator looking on at his subject, in effect. Moving in from the right is Mercury, the messenger of the gods: all smiles, this androgynous figure contrasts greatly with the countenance of the two other protagonists.

Van Baburen depicts Prometheus before his violation by the eagle. Some 300 years later, the Austrian Oskar Kokoschka (1886–1980), in contrast, would locate Prometheus firmly in the anguish of his adversity (and understandably so). Badly wounded during the First World War, Kokoschka had taught at the Dresden Academy from 1919. Declared by the Nazis to be a 'degenerate', he moved to London in the late 1930s.[24] Three years before his departure for Switzerland in 1953, he painted *Prometheus* for the right-hand panel in a *Prometheus Triptych* commissioned by the Austrian collector, Count Antoine Seilern, to decorate the ceiling of one of the rooms in his South Kensington house in London.[25] On the left were *Hades and Persephone*. The centre panel, *Cupid and Psyche*, an autonomous work completed in 1955, was seriously damaged in transit.[26] Executed in Kokoschka's characteristic Expressionist style, *Prometheus* is all pigment and texture and clashes of reds and yellows and fleshy pinks (illus. 31). Prometheus is shown shackled and writhing. He is a victim of torture, no more, no less. The eagle places its left foot on the Titan's thigh, all the while looking eagerly at his groin, which is a muddled mash of paint (at the bottom right a nonplussed bird looks on). Kokoschka's *Prometheus* is a reminder of all tortures. It goes beyond this, however, and involves, disturbingly, the sexualization of its subject; something which now resonates strongly in the light of the obscenities US soldiers committed on Iraqi prisoners at Abu Ghraib Prison (2004–6).[27]

Some 23 years before Kokoschka painted Prometheus, the great Mexican muralist José Clemente Orozco (1883–1949) had visualized a Prometheus whose otherwise mighty masculinity is compromised by his (apparent) lack of genitals. The first Mexican fresco to be completed in the US, *Prometheus* is the massive, central panel in the refectory of Ponoma College, Claremont, California (illus. 32).[28] The west panel shows *Zeus, Hera and Lo* or *The Destruction of Mythology*; the east panel, *Centaurs in Agony* or *Strangulation of Mythology*. On the ceiling, is *Abstract Composition* or *Godhead, Prometheus*. Set against a blue background, it is a stark arrangement of rectangles and stylized flames shooting off to the left and right.

Orozco's nude Prometheus towers above Dantesque figures: those on his right supplicate or embrace; a few turn their backs on the Titan. To his left, they bow their heads or throw their hands up in dismay and/or awe.[29] Here, the giving of fire is seen as a mixed gift. So too is it in the east panel, where centaurs are represented either yielding to or caught in the coils of an enormous serpent; a baby centaur, bottom left, lifts its arms in a pathetic gesture. According to Orozco, this panel represented 'the ancient times that Prometheus is upsetting by giving knowledge to man'.[30]

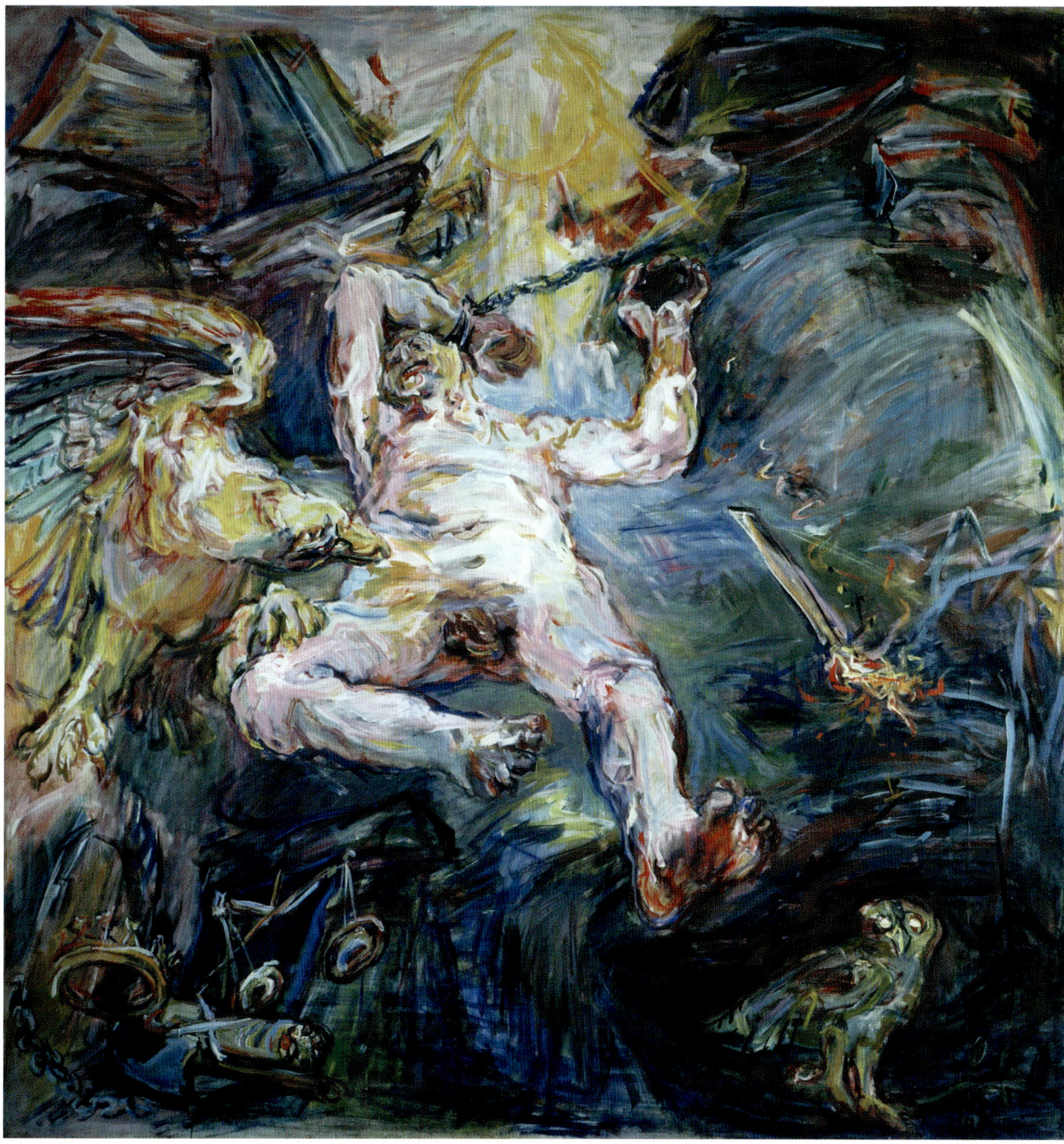

31 Oskar Kokoschka, *Triptych – Prometheus*, 1950, tempera on canvas.

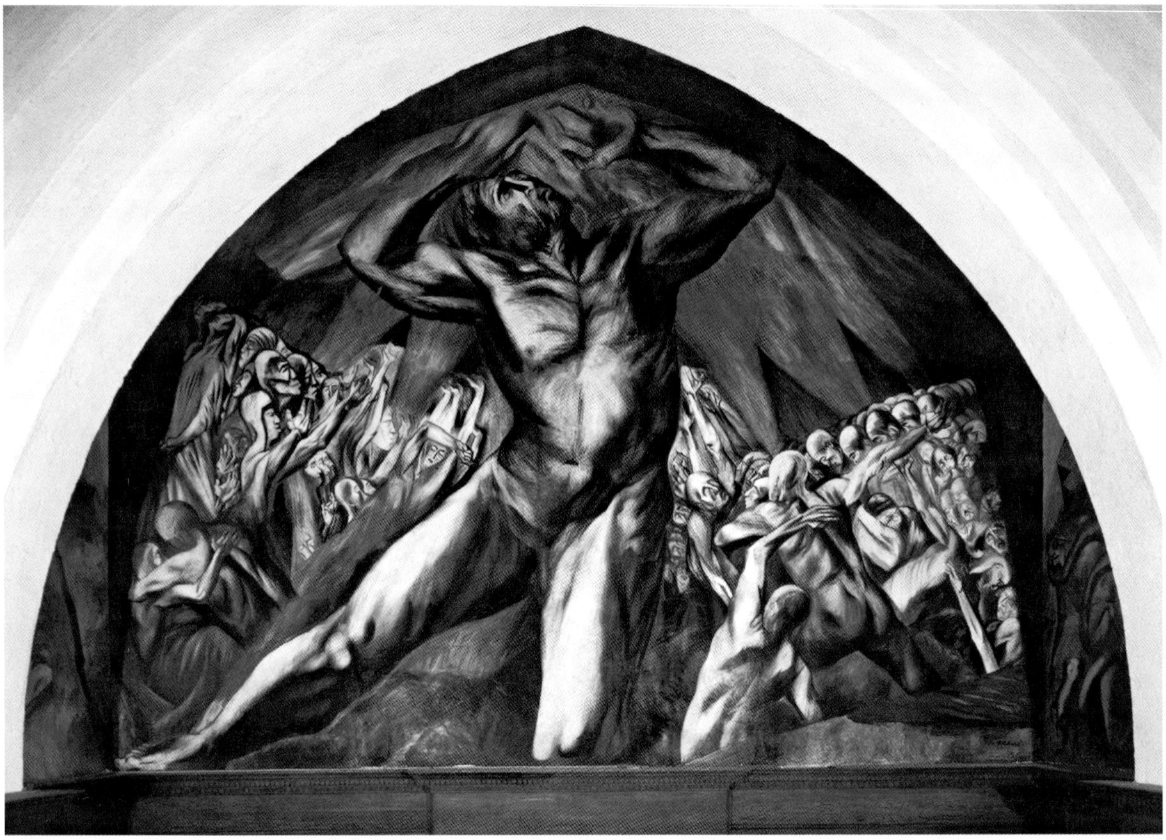

32 José Clemente Orozco, *Prometheus*, 1930, central panel from a fresco mural at Frary Hall, Pomona College Museum of Art, Claremont, California.

'Knowledge' here, of course, is fire itself. On the west wall, Orozco shows Zeus and Hera gazing agonizingly up at Prometheus; Lo, hybrid animal and human, and the third protagonist, is both beseeching and calm. Once again, Orozco draws attention to the ambivalent nature of fire. Which takes us back to his representation of a sexually ambiguous Prometheus in the central panel: 'ambiguous' because, while much has been made of Prometheus' absent genitalia, they are in fact there – a forlorn little penis pointing down between his thighs. This is an *übermensch* (a superman) brought down to earth.

Dancing and Damnation

These unstable representations of Prometheus invite comparison with two sculptures that bring fire's contradictory symbolism into sharp focus. Some 100 years separate these finely crafted, smallish objects: one is a central image in Hinduism showing *Siva Nataraja (Dancing Siva)* (illus. 33); the other, less well known, is by Francesco Antonio Picano (*d.* 1743) and Lorenzo Vaccaro (1655–1706), and depicts *St Michael Casting Satan into Hell* (illus. 34).

Flames gesticulate around the four-handed, contorted figure of Siva while, in the richly coloured (hyperreal) Pizano/Vaccaro, flames brush Satan's flesh. Reptilian-like, the latter speak of fire's intimate links with the corporeal. Dante's own vision of Hell in the famous *Inferno* (the first volume of *The Divine Comedy*) recognized as much: it is one of a netherworld of bodies, some contorted and mutilated (Cantos 24 and 28), some diseased and tormented (Canto 29), and others grossly transfigured (Canto 13).[31] In Canto 9, where the poet with the help of an angel enters the city of Dis, he finds the heretics in sepulchres burning with fire: 'Their [the tombs'] lids were all ajar, and out of each / came groans so terrible they seemed the cries / of spirits mad with suffering and grief.'[32]

The tortured expression of Satan in the Picano/Vaccaro is the antithesis of St Michael's effete countenance. Beelzebub is naked and pink and fleshy whereas the Archangel warrior, responsible for the defeat of Satan and his band, is resplendent in colourful wings and flowing garments. He raises his sword in a gesture that is never to be realized; caught in a balletic pose, the saint lingers above Satan, interminably. This same (seemingly) eternal quality distinguishes the sculpture of Siva Nataraja. 'The cosmic dance of Siva', writes Basil Gray,

> is at its roots, the manifestation of primal, rhythmic energy . . . In a delicately balanced composition, the divine dancer moves with

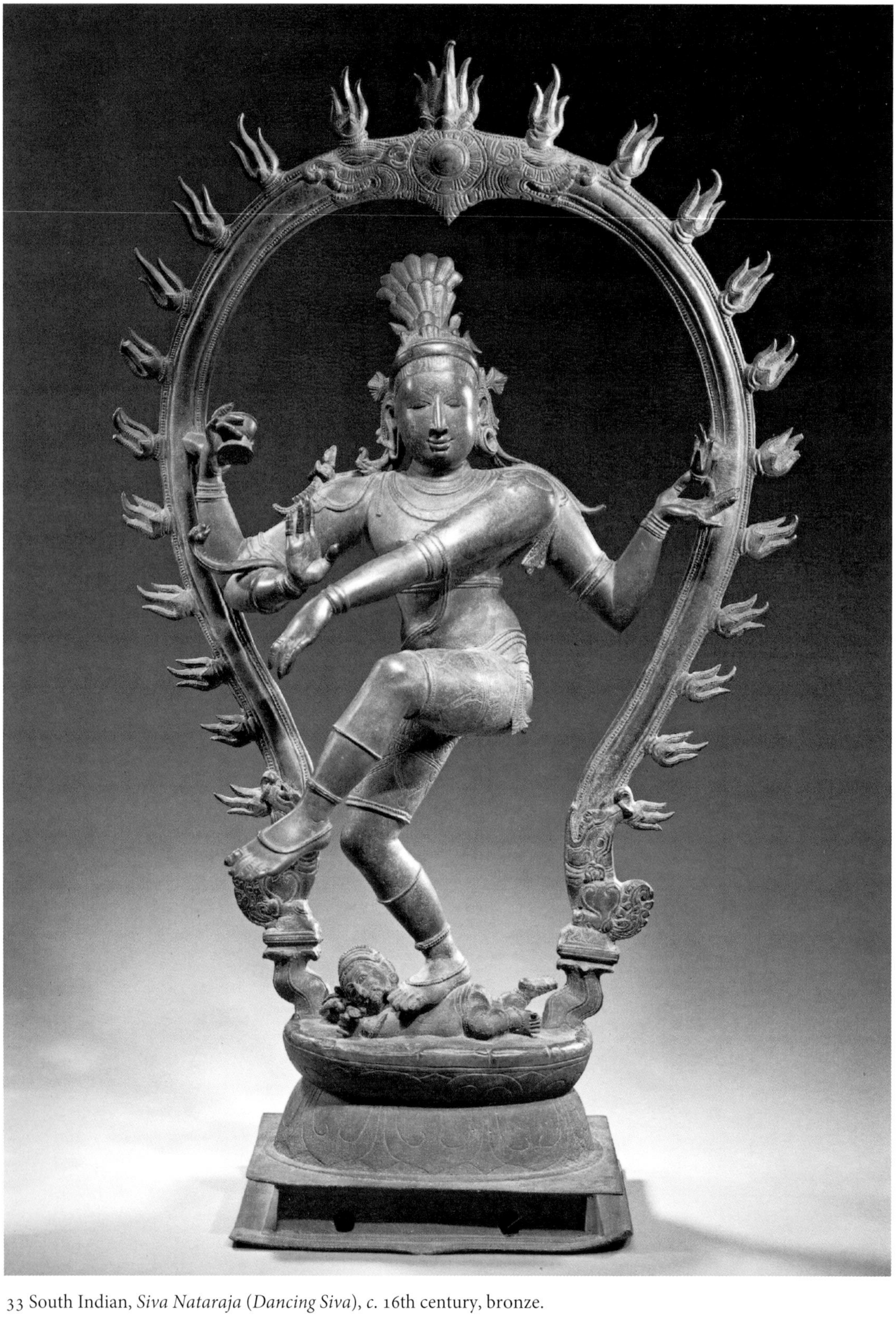

33 South Indian, *Siva Nataraja* (*Dancing Siva*), *c.* 16th century, bronze.

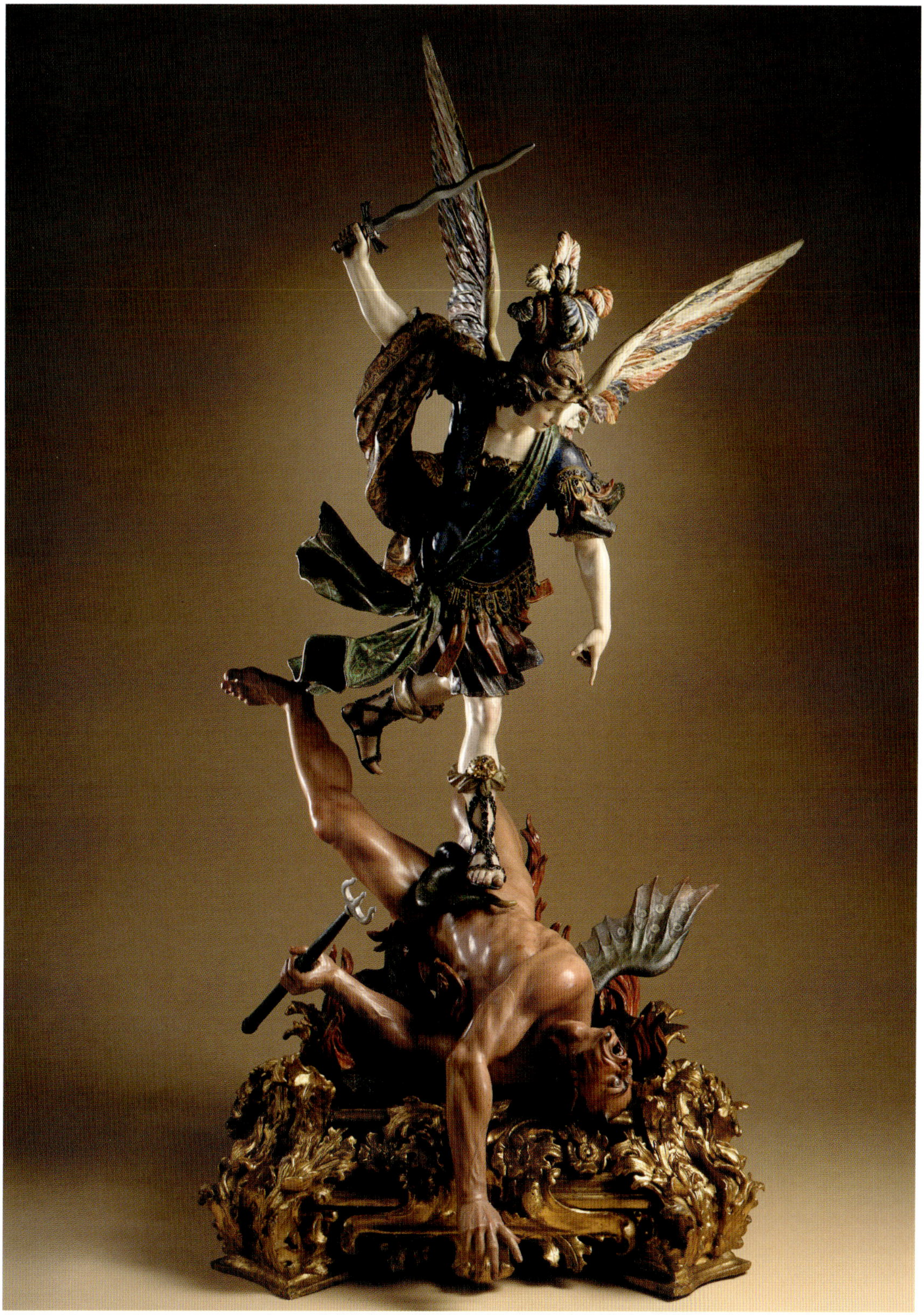

34 Francesco Antonio Picano and Lorenzo Vaccaro, *St Michael Casting Satan into Hell*, 1705, polychromed wood with glass.

> effortless grace . . . suspended in the air, in a magically arrested movement full of rhythm and grace.[33]

It is certainly this, but it is also much more.

Dancing Siva has a long history in poetry and literature, but its representations in bronze seem only to go back to the ninth century.[34] Mapping its interpretive possibilities in his authoritative book, *The Dance of Siva*, Ananda K. Coomaraswamy draws attention to three dances.[35] The first, an evening dance in the Himalayas, involves a two-handed Siva surrounded by a range of gods in the role of chorus; absent is the evil dwarf, Muyalka, trodden on by Siva, and seen in the work illustrated here. The second, more familiar dance, is the *Tandava*,

> performed in cemeteries and burning grounds, where Siva, usually in ten-armed form, dances wildly with Devi [or Mahadevi, 'The Great Goddess'] accompanied by troops of capering imps . . . Siva is a destroyer and loves the burning ground. But what does he destroy? Not merely the heavens and earth at the close of a world-cycle, but the fetters that bind each separate soul.[36]

Here, Siva demolishes and releases; the 'burning ground', in effect, becomes a redemptive site.

Coomaraswamy's third example relates directly to our example: Siva, four armed, stomps on a dwarf whose presence derives from the myth of the 10,000 Rishis, or sages. Determined to resist Siva's attempts to co-opt them to his cause, the Rishis responded by sending animals to devour him (first a tiger, then a serpent). When this failed, they sent a wicked dwarf brandishing a club to attack the god.[37] Siva, however, simply stood on the dwarf and commenced his wondrous dance: the sages were transfixed, and threw themselves at Siva's feet. Siva's dance, like fire itself, takes on different meanings and consequences. He dances out the creation of the world; he rebuffs opponents; and, in those moments of respite brought about by fatigue, universal chaos results.

Siva's gyrations take place in a framework of flickering flames; he also holds a flame in one of his outstretched hands.[38] This is appropriate, since fire holds a supreme place in Hinduism.[39] It is personified in the god Agni who, especially in the early formulations of Hinduism, was considered to be the supreme deity, with the exception of Indra, the great god of storms and war. It is in the Rig Veda, the ancient Indian sacred collection of hymns, that fire's multiple meanings are best explained. Some of these verses

> speak of fire's awful anger, and describe how it shears off great forests as if with a razor. But this aspect became submerged in the general feeling of the beneficent utility of fire as a means of production.[40]

The Kausitaki Brahmana of the Rig Veda (1.1) puts it thus:

> The gods said, 'The dread forms of thee as such we deposit apart, and then thou be here for man with that form of thine which is auspicious, helpful, and worthy of sacrifice'. Then in the waters they placed his blowing form, in the wind the purifying, in the sun the bright. Then with his auspicious, helpful, and sacrificial forms he burnt here for men. These indeed are the forms of Agni.[41]

This understanding of fire helps explain its strong presence in *Siva as Lord of Dance*, where it brings together the joyful and the dangerous, whereas in Picano's and Vaccaro's *St Michael Casting Satan into Hell*, it is lodged firmly in an ethos of dread and damnation.

Burning Bushes, a Babylonian King and Arthur Boyd

The two sculptures discussed above weave together narratives of fire's promise and blight, whereas in the story of the Burning Bush, from Exodus in the Old Testament, fire is all about the shocking and the revelatory. The story tells of how

> the angel of the LORD appeared unto him [Moses] in a flame of fire out of the midst of a bush: and he looked, and, behold, the bush burned with fire, and the bush *was* not consumed. And Moses said, I will now turn aside, and see this great sight, why the bush is not burnt. And when the LORD saw that he turned aside to see, GOD called unto him out of the midst of the bush, and said, Moses, Moses. And he said, Here *am* I.[42]

The Flemish artist Simon Bening (*c.* 1483–1561) illustrates this extraordinary event in a miniature that shows Moses at the bottom of the page removing his shoes and gazing at God in the holy bush above (illus. 35). Moses is all innocence and awe (Christians would come to see this event as a foreshadowing of Mary's conception of Christ without losing her virginity). Many years later, Domenichino (1581–1641) would similarly depict a star-struck shepherd in *Landscape with Moses and the Burning Bush* (1610–16). Here, though, the figure raises his hand to shield himself against the incredible occurrence.

A different perspective on the Burning Bush is revealed in the iconography of the Orthodox Church in late Imperial Russia, where an image of the Mother of God, who was believed to be the guardian of the home against fire, would often be hung in a corner opposite the stove in the domestic home.[43] Usually comprising the figures of Mary and the Christ Child set against a four-pointed medallion, some of whose points were coloured red, this relationship between icon and stove, as Cathy Frierson has observed, found expression in a peasant saying popular in

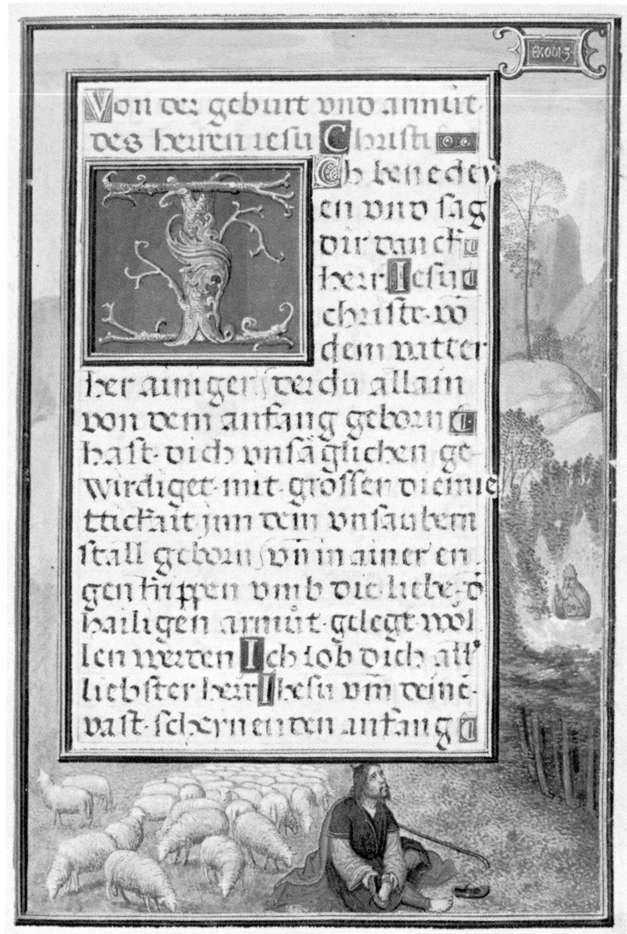

35 Simon Bening, '*Moses and the Burning Bush*', miniature from the Prayer Book of Cardinal Albrecht of Brandenburg, folio 20, *c.* 1525–30, tempera, gold paint and gold leaf on parchment.

Viatka Province: 'A stove in the house is the same as an altar in church: bread is baked in it.'[44]

A variation on the theme of the Burning Bush resonates in another Old Testament narrative (from the Book of Daniel) of the three young Jewish men, Shadrach, Meshach and Abed-nego, who refused to worship before the image of gold made by Nebuchadnezzar, King of Babylon.[45] This was illustrated, for example, by the celebrated English artist J.M.W. Turner (1775–1851), in a painting he showed at the Royal Academy in 1832. Nebuchadnezzar had ordered that the recalcitrants be bound with ropes and thrown into a blazing furnace. The intense heat killed those who opened the furnace to cast in the fractious trio, while the latter miraculously survived. Overwhelmed, Nebuchadnezzar now heaped praise on the men of God; thereafter, they lived a prosperous life in Babylon at the King's pleasure.

One of the most striking images of Nebuchadnezzar is that by the English artist William Blake (1757–1827), where he shows the king, ignominiously humbled by God for the hubris he displayed in the building of Babylon, crawling on all fours like an animal.[46] Blake's interest in Nebuchadnezzar extended over some fifteen years, from 1790 to 1805. Much later, the Australian artist Arthur Boyd (1924–1999) would similarly be attracted to Nebuchadnezzar's troubled and tantalizing history. Collaborating with the Oxford art historian, T.S.R. Boase, whom Boyd had met when that scholar visited Australia in 1956, the artist had 34 paintings as well as prints and drawings on the theme reproduced in Boase's book *Nebuchadnezzar* (1972).[47] Claimed in that text to have been executed between 1968 and 1971, it is possible that some of these paintings were started as early as 1966. It is now also clear that Boyd had produced more variations on the theme than those Boase had included.[48] One of these, *Nebuchadnezzar in a Fire* (illus. 36), draws freely on Blake's own image of the king on his hands and knees (Boyd was often influenced by historical art, including works by Bosch and Breughel). Now reduced to a white, whispery figure, Boyd's Nebuchadnezzar crawls through a firestorm: open-mouthed, he appears to gulp something black and coiled-like. Or is he uttering some primeval scream? There is no fire in the Blake. Nor is one mentioned in the Book of Daniel raining down on the Babylonian king. Fire here, as in Boyd's other works in the series, is an invented motif; there is one exception, *The Fiery Furnace*,[49] which depicts Shadrach, Meshach and Abed-nego in the midst of the flames, but even in this painting Boyd parts company with the biblical narrative: his young men are naked rather than 'bound in their coats, their hosen, and their hats, and their *other* garments'.[50] In *Nebuchadnezzar on Fire*

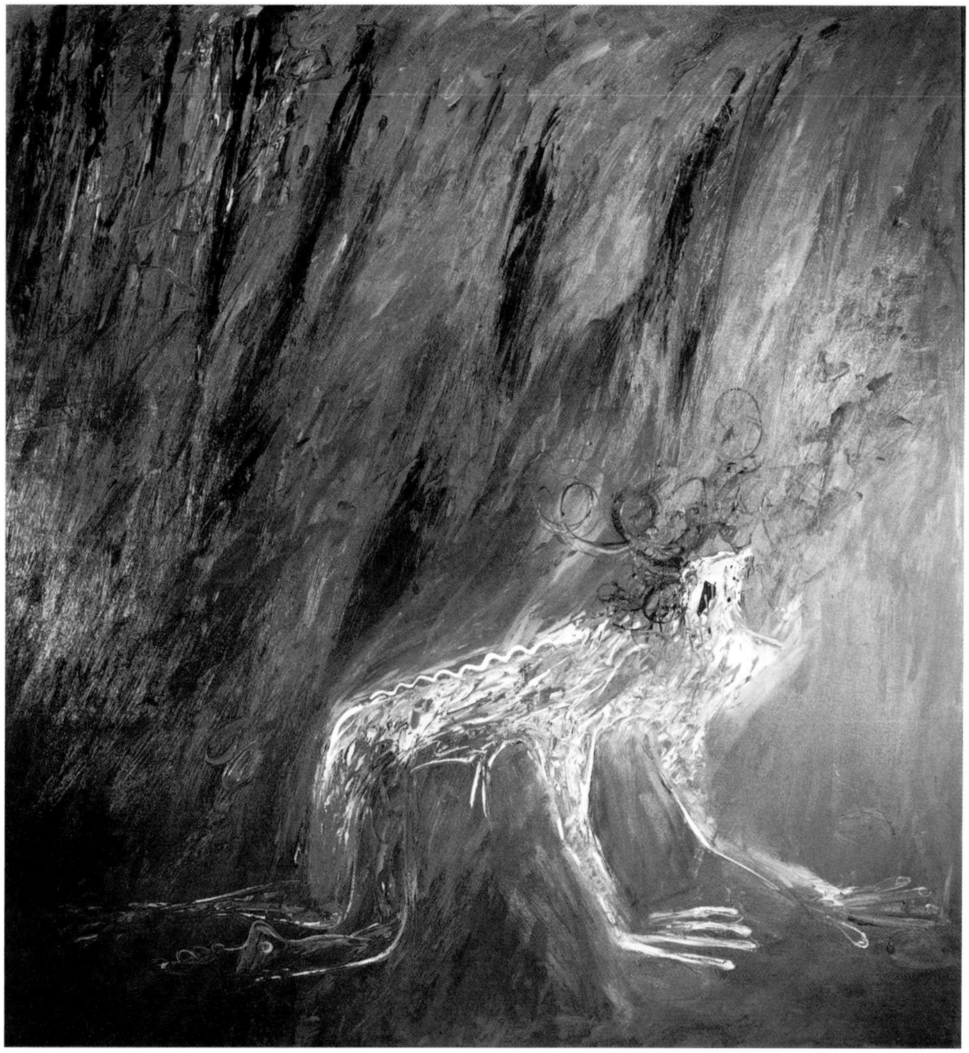

36 Arthur Boyd, *Nebuchadnezzar in a Fire*, *c.* 1969, oil on canvas.

Falling over a Waterfall (illus. 37), and in *Red Nebuchadnezzar with Crows* (plate six in Boase's *Nebuchadnezzar*), to cite just two other examples, fire is ubiquitous. The former presents a blazing king diving headlong into a cascade set in a wretched landscape of tentacle-like trees and roots. Arms apart and his face a picture of grim determination, not unlike that

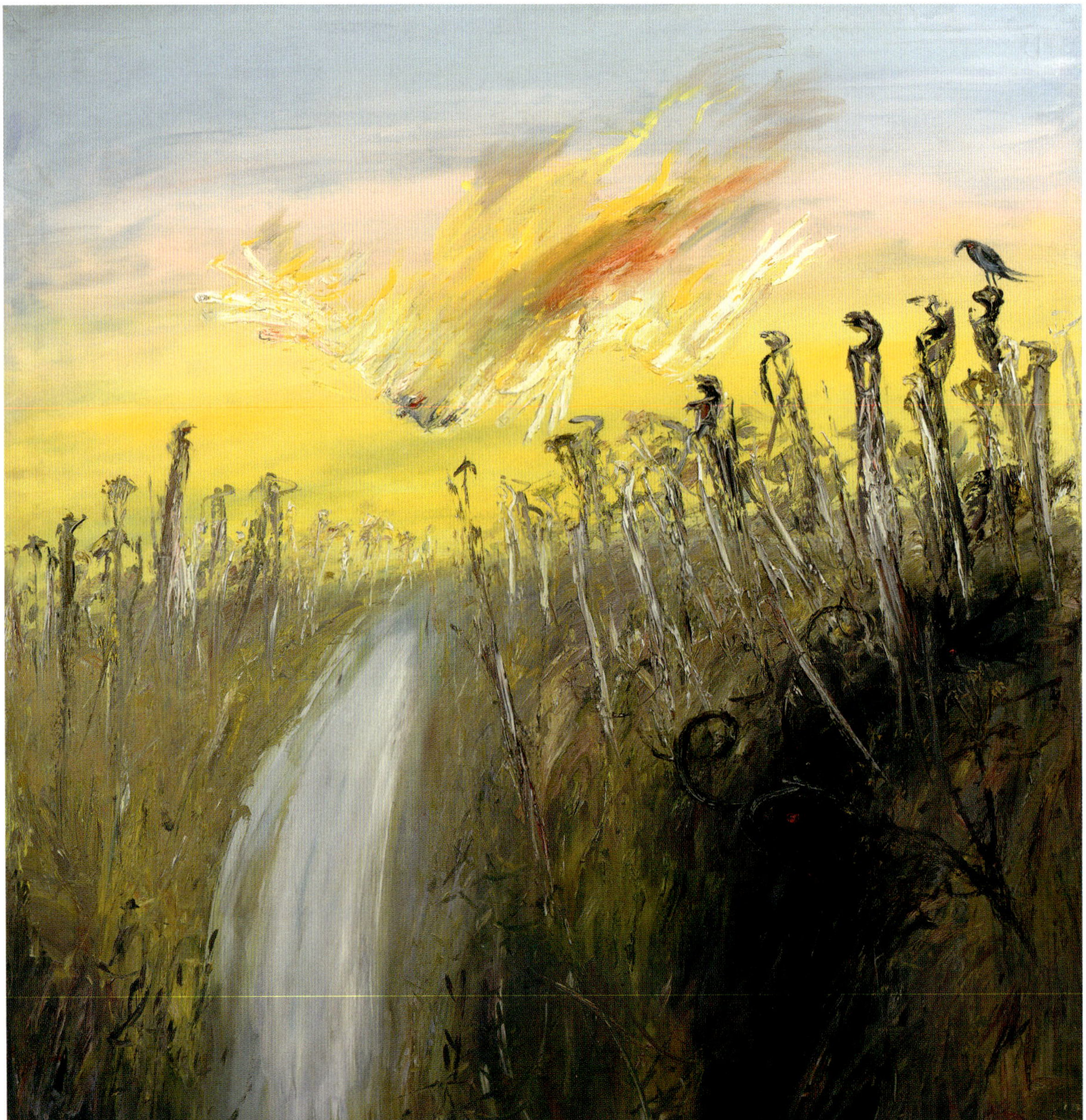

37 Arthur Boyd, *Nebuchadnezzar on Fire Falling over a Waterfall*, 1966–8, oil on canvas.

of a gritty skydiver, Boase describes this Nebuchadnezzar as a 'flaming meteor, and the waterfall that Boyd has imagined below him cannot quench it': an inversion of the polarities of fire and water, where the latter is typically framed as ascendant.[51] Boyd's image, of course, also brings to mind the celebrated tale of Icarus who, together with his father, Daedalus, escaped from the Labyrinth by attaching wings to their shoulders with wax and then flying away.[52] Icarus flew too close to the sun, however, which melted the wax and he fell into the sea and drowned.

It is appropriate here to cite the New Zealand artist Euan Macleod (*b.* 1956), whose painting *Exquisite Corpse with Fire* (illus. 38) also echoes the Icarus saga. An intensely visceral work, it takes its title from the French term *cadavre exquis*, which the Surrealists employed to describe this simple parlour game of chance that they made their own. Macleod shows an inverted figure of a man, arms outstretched, whose torso is made up of red-hot paint thickly applied. Hovering above the figure is what appears to be a swirling pool of water. Where the man's head is, at the base of the painting, there is deep-blue sky. It's a topsy-turvy picture evoking an overriding sense of vertigo that brings strongly to mind Icarus and his fatal, dizzying escapade.[53]

Meanwhile, in Boyd's *Red Nebuchadnezzar with Crows*, all is scorching red and slashes of yellow. His body in flames, the king is depicted as a disorientated figure moving awkwardly amongst dense undergrowth. Two crows with red eyes fly towards the king like lethal warplanes. These creatures are distant reminders of Nebuchadnezzar's disturbing dream related to Daniel, in which the king says he saw 'a tree in the midst of the earth' reaching towards heaven:

> The beasts of the field had shadow under it, and the fowls of the heaven dwelt in the boughs thereof . . . Hew down the tree, [a 'holy one' said to Nebuchadnezzar] and cut off his branches, shake off his leaves, and scatter his fruit: let the beasts get away from under it, and the fowls from his branches.[54]

Daniel would interpret this dream as a portent of the king's demise, something that Boyd depicts but with his characteristic license.

Boyd's preoccupation with fire and destruction is consistent with the narratives of good and evil, redemption and ruin evidenced in earlier works such as the series that eventually became known as *Love, Marriage and Death of a Half-caste* (1957–59). As the art historian Ursula Hoff has pointed out, however, the Nebuchadnezzar paintings were executed at the height of the Vietnam War, where images and commentary on burnt and burning villages, 'children set on fire by napalm', were commonplace.[55] Barry Pearce, who organized the 1994 Boyd retrospective at the Art Gallery of New South Wales, Sydney, builds on this idea by suggesting that self-immolation may also have provided the artist with a reference point. Pearce writes:

> Self-immolation in protest actually took place on Hampstead Heath near Boyd's home, and once more a biblical theme was seen to be an allegory of the descent of humanity in a conflicted world.[56]

Boyd and his family had left Australia for England in 1959 and, as he now lived in Hampstead, he would surely have known of this incident of 'self-immolation' enacted by a man on the Heath on 26 April 1967; only later was the disfigured body identified as that of the Irish author Peter Tyrrell. Contrary to what Pearce implies, however, Tyrrell's death (Pearce never identifies him by name) had nothing to do with Vietnam; it was a suicide brought about by dire personal and political circumstances.[57]

One suspects that for Boyd, Tyrrell's death by fire resonated rather with an awareness of the wider instances of self-immolation that had characterized the early years of the Vietnam conflict. Certainly the story of Thich Quang Duc, the 73-year-old Buddhist monk who set fire to himself in Saigon on 11 June 1963, would soon fill the pages of the international press: *Life* magazine would carry images of protesting

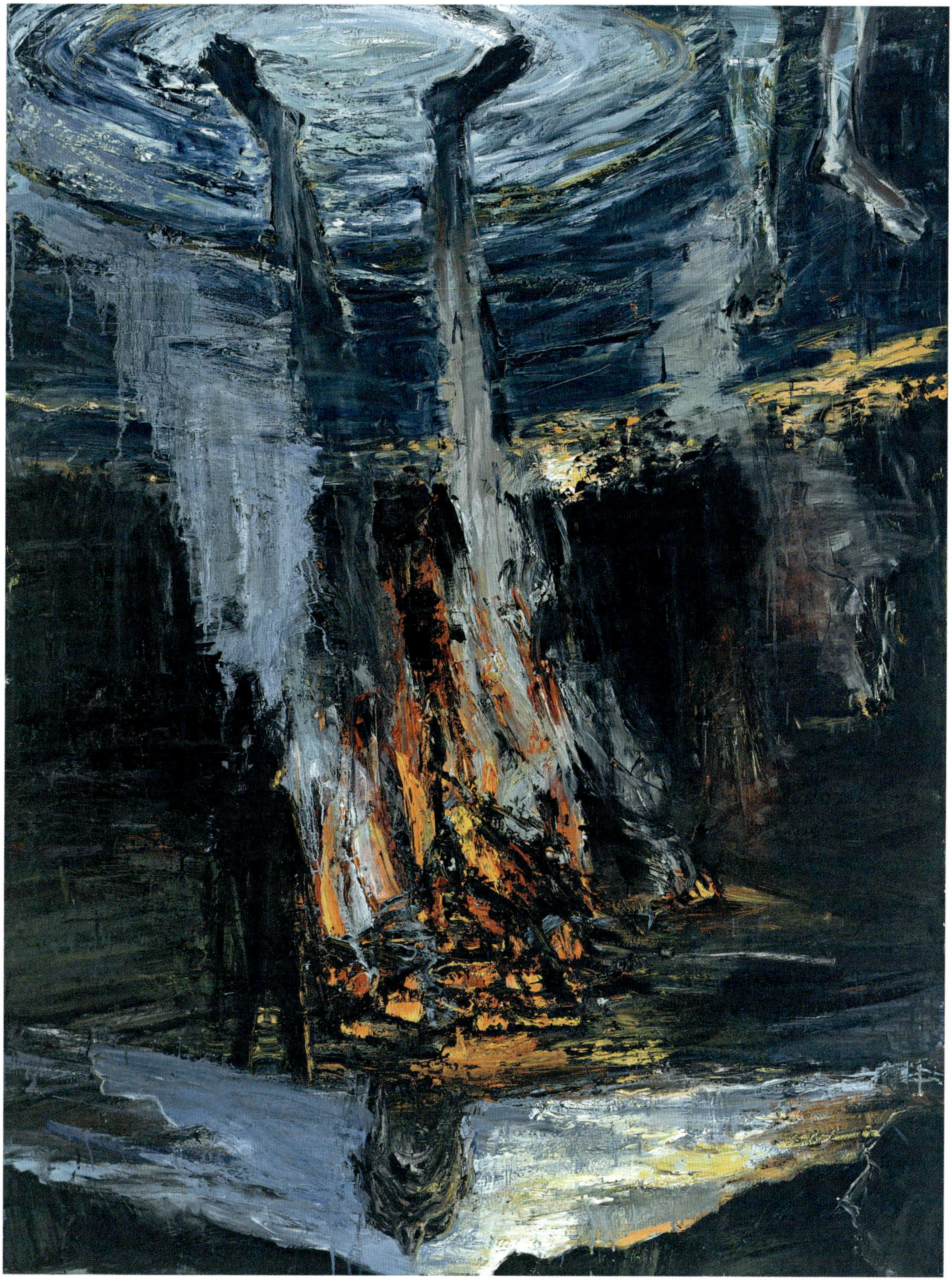

38 Euan Macleod, *Exquisite Corpse with Fire*, 2000–1, oil on canvas.

or burning monks in its June, August, September and November editions of 1963 (illus. 39).[58] Writing in the *New York Times*, David Halberstam spoke about the burning of Quang Duc's body – 'slowly, witheringly and shriveling up, his head blackening and charring'.[59] These, of course, are the more obvious – and ghastly – effects of fire on a body. But as it turned out, surprisingly Quang Duc's heart had not burnt, and it was recovered and treated as a sacred artefact.

Most commentators understood Quang Duc's death to be essentially a religious matter, a case of self-sacrifice: a protest against the persecution of the Buddhists in South Vietnam by the Vietnamese Roman Catholics. Self-immolation was seen to be consistent with Buddhist belief; an action, in the case of Quang Duc, that was considered to be quite unlike those of 'their Western imitators . . . [in that it does not] constitute purely political protest actions'.[60] This view would be questioned a few years later by

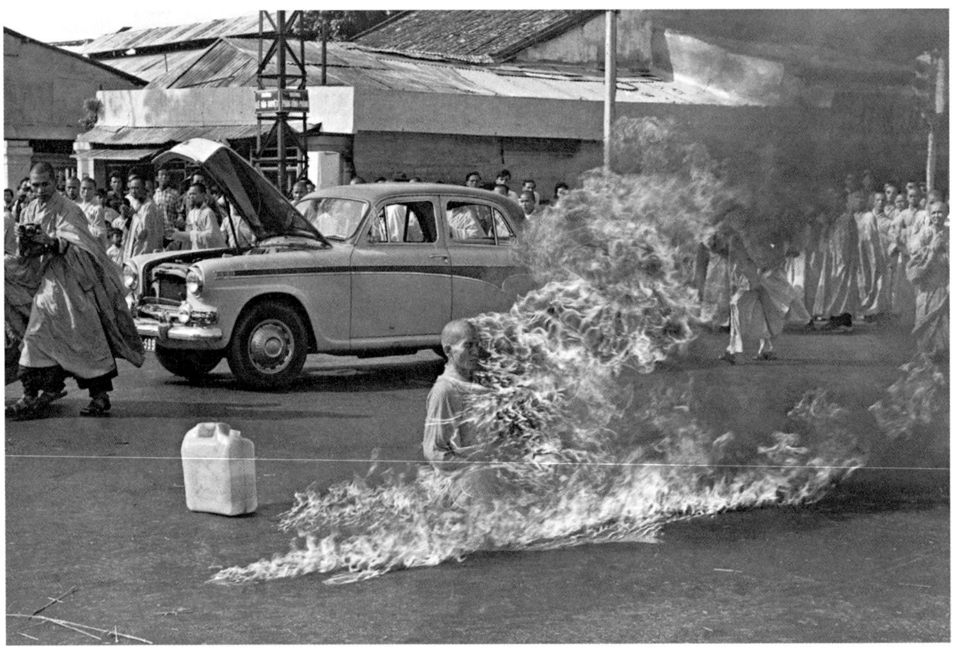

39 A Buddhist monk burns himself to death in a protest against the puppet Diem regime, photograph by Malcolm Browne, 11 June 1963.

the Vietnamese Zen Buddhist monk, writer and poet Thich Nhat Hanh, who argued that Quang Duc's self-immolation should be understood not simply in religious terms but within its broader social, economic and political contexts.[61] Emphasizing that, since the introduction of Roman Catholicism in Vietnam in the sixteenth and seventeenth centuries, Roman Catholicism had always been intimately associated with white explorers, missionaries and the mercantile class, Nhat Hanh makes a convincing case for seeing Quang Duc's suicide as more a political than sacrificial or religious event.

Three Festive Fires

When united fire and faith can be explosive. And so it is, literally, with the Las Fallas festival in Valencia, Spain, which commences on 1 March and concludes at midnight on St Joseph's Day (19 March), with the burning of intricate, huge sculptures and effigies, known as *fallas* (illus. 40), accompanied by fireworks. Many thousands of visitors and locals take over the streets and assemble in main squares to celebrate this colossal public burning.[62]

The word *falla* derives from the Latin *facula*, meaning 'torch'; these torches were employed by Valencian carpenters when working during the long winter nights. Come spring, the sculptures were burnt to celebrate the feast day of St Joseph, the carpenter and Christ's earthly father. Effigies, it seems, were then placed on the bonfires: some of these now take on the attributes of well-known politicians, while others form part of elaborate tableaux commenting humorously but pointedly on local and international issues.

The social satire and caricature evident in these constructions embody the notion of 'criticism' that, as Xavier Costa has made clear, is closely related to cleansing and renewal through fire – something that

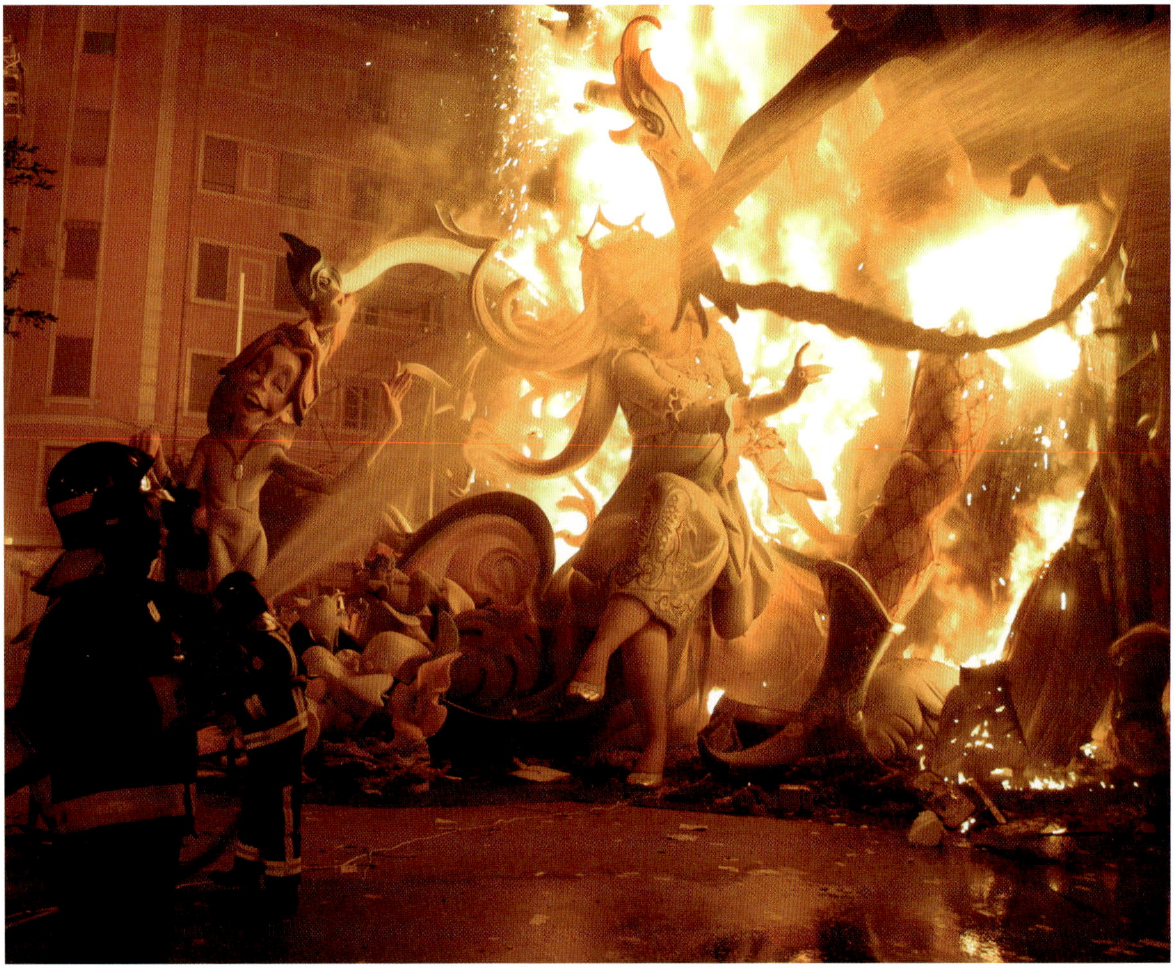

40 Flames engulf puppets as firefighters hose down a fire as the city of Valencia celebrates the finale of its 'Fallas' festival.

is strongly bound up with the legendary sacred firebird, the phoenix. An artist involved in the making of a *falla* put it thus:

> The Falla begins with a critical sense of humour. The content of this critique is social, economic and political. The meaning of it is that the Falla is erected, exposed to the public and burnt, to be immediately born again. There is something fantastic here. And some

> people don't understand why it has to be burnt . . . It is necessary to burn things. This is not like the activity of the banks that keep things. No! no! We set things on fire so that other things can grow again from the ashes. This is the meaning of the Fallas, and I find it beautiful.[63]

This transient beauty of fire is also symbolized in the flowers that form a ubiquitous part of the festival. Ephemeral like fire itself, flowers are given to the *fallera majors* (the festival queens) and the Virgin, but the participants themselves may also keep them.

Simon Wallis gives this vivid account of his first encounter with a *falla* while following a procession through the Costa Blanca town of Gandia:

> Towering over thirty feet, looming as if about to pounce, was the soulful bearded face of a wizened old man . . . To one side was Old Spain; grey men, shrouded in cobwebs . . . On the other rose an image of Gandia's latest park, a dubious development including high rise flats, crowned by a Burger King . . . Small issues, global social trends, and the mortality of man juxtaposed with the infinite quality of time were all encompassed in one gaudy, magnificently monstrous sculpture.

Later on Wallis describes a *falla*'s fiery climax:

> I thought the fire had not caught hold. Then I realised there was burning from the inside [of the *falla*]. With a 'crack' one of the supports gave way and a ball of fire belched through the gap . . . Around it people clamped their hands over their foreheads to protect the expanse of bare flesh. That night I watched three *fallas* rise in flame and fall in ashes.[64]

This description of a communal burning relocates some of fire's grim signifiers – belching, flesh and conflagration – in a context that, by contrast, is convivial and animated, one that invites comparison with the Naga Fireballs of Nong Khai, an annual festival held in October on the banks of the Mekong River in Thailand. Known by locals as *Bung Fai Phya Nak*, this event takes place during the full moon on the final day of Lent. Now a major tourist attraction, it is described in one of the festival's many websites: 'small balls of fire rise from the Mekong River and plunge upwards of 300 meters into the night sky. The fireballs appear as eerie blobs of red, blue, pink or green light that hang in the air for around 10 seconds.'[65] Simple and to the point, this description belies a more nuanced and controversial phenomenon. According to Buddhist beliefs, Nagas are semi-divine, water deities (serpents) who guard the religion's texts. Considered by believers to be the breath of the Naga, the fireballs have been described by a (initially) sceptical Gary Walsh:

> At 6.39 pm precisely, three fireballs rose in quick succession from the Laos side of the river . . . This is a swizz, I thought . . . I imagined a fireworks crew giggling as they set off skyrockets between the trees . . . And then came the moment that turned a doubter into a fervent believer. From directly in front of me, from the middle of the river, two fireballs rose in succession and flew swiftly over my head. The crowd cheered. I cheered. The Naga lived.[66]

The fireballs are there, indisputably, but where do they come from? One theory is that they are a natural phenomenon brought about by methane gas bubbling up from the riverbed, while a Thai TV station, ITV, claimed in a documentary some years ago that they were tracer rounds from AK-47s fired by soldiers across the Mekong River.[67] Natural, made by human hand or a miracle, or, in some strange ways a combination of all three, these claims were central to the film *Making Full Moon Party* (*Siop-ha*

Kham Duean Sipet), directed by Jira Malikul and released in October 2002 just days before the actual Naga Fireballs. For Malikul, it is precisely the contesting discourses surrounding the event that are critical: 'Everyone has his or her own version of the Naga Fireballs. This makes the event so special.'[68] The plurality of voices that Malikul invokes in the film – including a Buddhist monk (Luang Pho Lo); a young orphan (Khan); a representative of science and the rationalist view of the Fireballs (Dr Norati); and female characters like Teacher Alice and Aunty Ong, who embody an enduring faith in the Fireballs but also act as mediators between tradition and modernity – all these voices, like fire itself with its multiple, contradictory readings, are the stuff of *Making Full Moon Party*.[69]

Phrakhru Pichai Kijaton, abbot of Wat Paa Luang, reminds us that the temple houses had recorded the sighting of fiery orbs hundreds of years ago. 'Are they real? Does it matter? Faith is the thing', he says.[70] And 'faith' in one form or the other drives the (increasingly popular) Burning Man festival, in which fire, unsurprisingly, speaks in many different tongues. A week-long so-called countercultural event held annually since 1986, Burning Man was the brainchild of Larry Harvey and his friend Jerry James. Interviewed by Anthony Haden-Guest, who gives a sympathetic account of Burning Man in the journal *Art in America*, Harvey says:

> What I always hated about hipster scenes and probably would have hated about any of the avant-garde movements is the exclusivity . . . My people were dust-bowlers . . . My father was a carpenter . . . I went to a suburban high school and embraced their values in a lot of ways but we were outsiders to say the least.[71]

Such experiences helped inform Harvey's early philosophy. There was also, he recalls, an 'acid trip one day' that would convince him radical change was not easily come by; in addition, he remembers fondly evenings of 'old-fashioned bohemia'. From these inauspicious beginnings came Burning Man. Harvey telephoned a friend and said: 'Let's build a figure. Let's build a man. Let's burn him on the beach.' The rest, as they say, is history.

The first burning was a simple affair held on a Los Angeles beach on 21 June 1986. By 2004, and now ensconced in the Black Rock Desert, Nevada, the festival was attracting some 35,000 visitors (illus. 41).[72] One year later, this figure had increased to 38,000. In 2008 it topped 49,599.[73] It is tempting to think that Harvey's idea for Burning Man was inspired by the huge success of the 1973 British cult movie, *The Wicker Man*, directed by Robin Hardy, which focuses on a fictitious Hebridean island called Summerisle where the inhabitants regularly participate in fertility rituals and the worship of the sun.[74] One of the film's chief characters, Sergeant Neil Howie, investigates the death of a young woman, Rowan Morrison. Howie is finally forced into a wicker statue of a man that is set on fire. Harvey says, however, that he had never seen the film; Burning Man, he stresses, began as a celebration of the summer solstice.[75]

In this respect, then, the festival picks up on many cultures that celebrate combinations of the winter and summer solstice, such as the ubiquitous Christian Christmas, the Jewish Hannukah, the Iranian Yalda and Kwanzaa, a celebration in the US that honours African history and culture and takes place in the last week of December. With one or two exceptions – Hannukah, for example, which commemorates the rededication of the Holy (second) Temple of Jerusalem in the second century BC with the kindling of candles in a nine-branched candelabrum known as a *menorah* – fire plays little or no role.

Not so in Burning Man, where it is everywhere. But the festival now sweeps across ideas and representations that range from neo-Dada installations and performance art to manifestos on self-sufficiency and environmental sensitivities: the 2007 festival was billed as 'The Green Man'. Amongst the motley array of artworks seen at Burning Man in recent years was a sci-fi figure hanging in a scaffold on wheels; two rather forlorn-looking rabbits with female 'drivers'; a massive head cut off just below the nose (Pepe Ozan's *The Dreamer*, 2005) and a

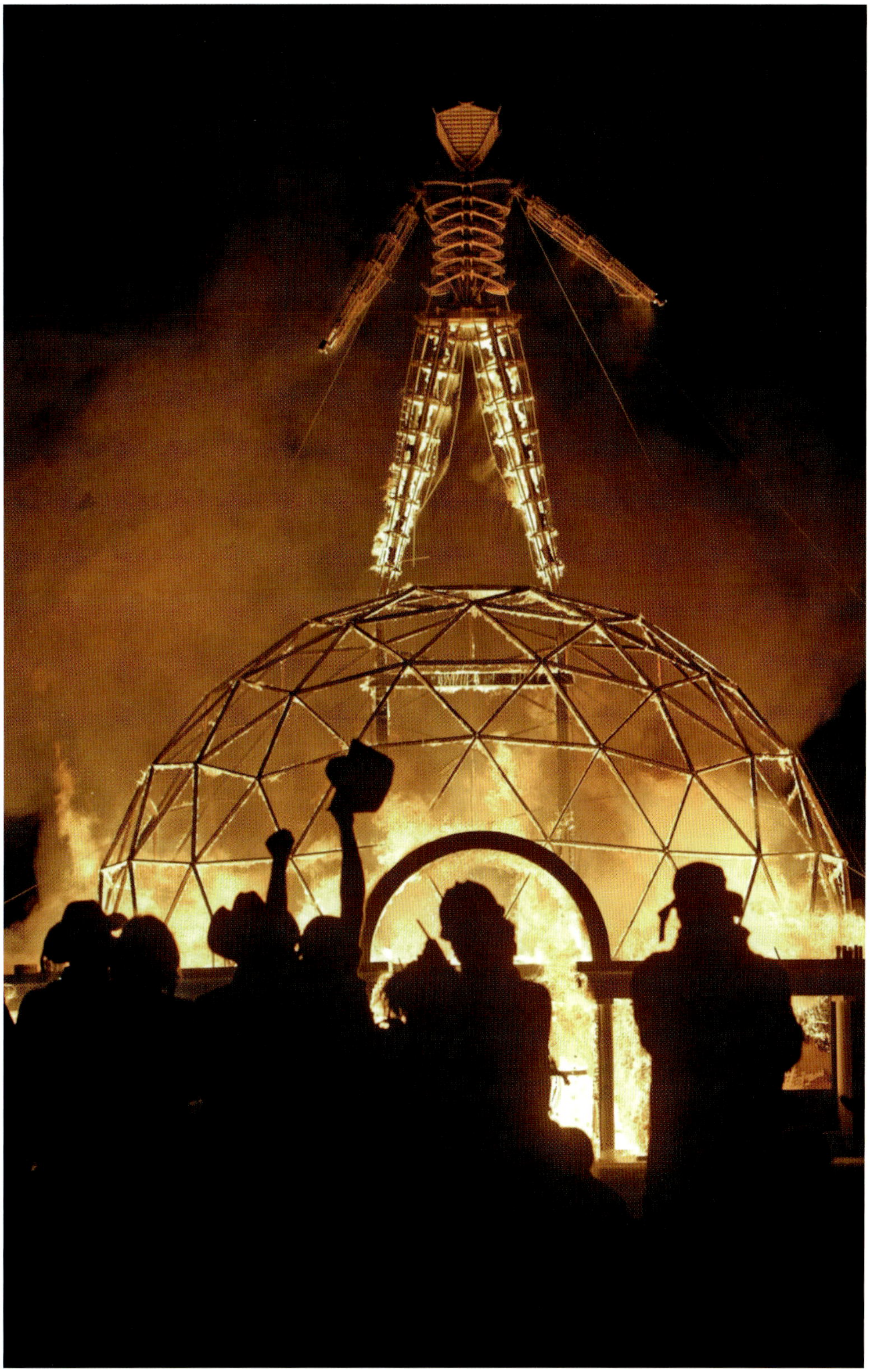

41 *Burning Man*, 2004.

performance involving two men hitting themselves with foam bats. At the heart of the event, though, is the burning of a 'man'. A 12-m (40-ft) neon and wooden structure was erected for this purpose in 2004 (see illus. 41). For those who stayed until the end of the festival, there was another major burning, this time of the 'Temple of Stars', created by David Best. Consisting of filigrees of plywood, the components were constructed in Petaluma, California, four months earlier, before being transported by trucks to their destination. Tradition requires that revellers insinuate the names of departed loved ones and other mementoes in the edifice. The *New York Times* reported:

> Many visitors cried while composing their gifts, and some collapsed into the arms of others. One middle-aged man read a lengthy, moving essay about his eccentric late mother, then fell to the floor of the temple and hyperventilated.[76]

For many, torching the temple has become the centrepiece of the event, a more intimate, spiritual moment than the rave-party-like immolation of the man icon. 'To burn it, it's like giving it a higher force', said Fred Dickson, who helped build the temple. 'It's like an offering.'[77]

Burning Man functions on a number of levels. For one thing, it is an opportunity to engage in jubilant self-expression. It is, of course, also about community engagement. There were precedents for all this in the rock festivals of the 1960s, which, like Burning Man, had a penchant for pronouncements of a mystical nature. But compared to these it is far more market-savvy: consider its very seductive website and its heightened sense of its own history. A more singular comparison with the 1960s suggests itself, but even this fails: at the famous Monterey Pop Festival held in California on 18 June 1967, Jimi Hendrix (1942–1970) concluded his memorable performance by setting his guitar alight and then destroying it.[78] This was a rebellious gesture, albeit one undertaken with an enthralled media in mind, but it has little to do with Burning

Man's aspirations to cathartic renewal through fire: a desire that at best is utopian, at least, ingenuous. Either way, Burning Man's veneration of fire can only be understood in terms of fire's long history which, as outlined in the introduction to this chapter, is all about paradox and permit.

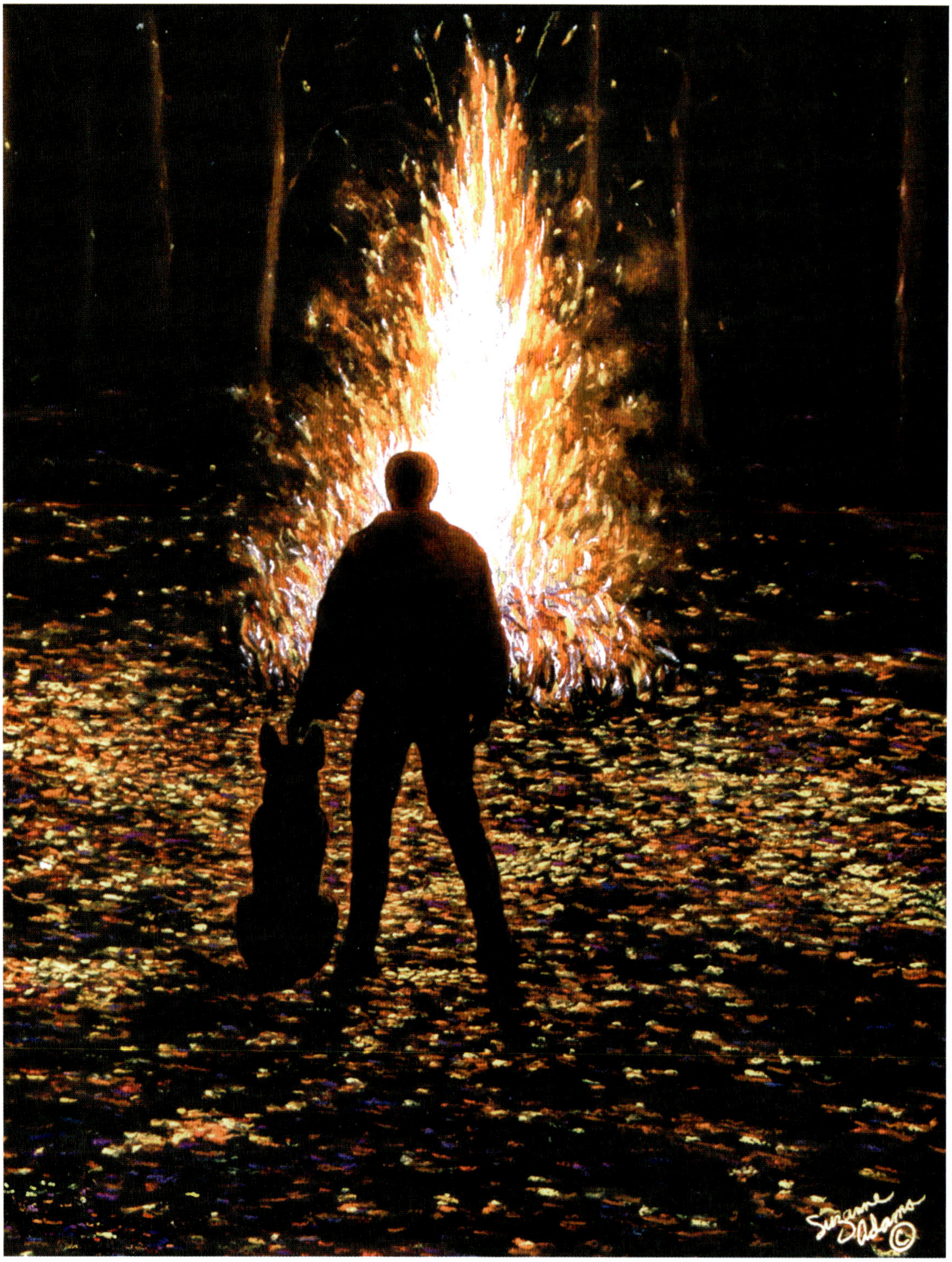

42 Suzanne Adams, *Fire Talk*, 1996, soft pastels.

three

The Intimate Flame

. . . unexpected pops, cracks, and sizzles, along with its unique, pleasant aroma.
Suzanne Adams[1]

Fire Talk, a painting by Suzanne Adams, shows a man and his dog, silhouetted in the foreground and seen from behind, staring at a bonfire (illus. 42). The figure stands with legs apart; his left hand barely touches the head of the dog (more precisely, it is placed gently between the animal's upright ears). Staring at the bonfire, and seemingly mesmerized by it, they both seem intimately related in their shared experience. Explaining her drawing, Adams mixes a quiet eloquence with an unbridled romanticism; she needs to be quoted at length:

> The painting was inspired by all the nights I watched my German Shepherd Dog by the fires I had in a clearing at the edge of my woods. 'Leben' (meaning 'lively' in German) was fascinated by fire and would dance all night, if you let him, catching the sparks that flew in the sky. Tom, my friend, shared the same enchantment, diligently adding and moving the wood to keep the fire alive and strong. Here they are at one of these fires in complete silence, fully absorbed in its changing, illuminating presence, unexpected pops, cracks, and sizzles, along with its unique, pleasant aroma . . . The two represent the symbiotic relationship man and dog have shared for eons of time, both in need of this ancient element for warmth and light. Only man and the powers of heaven can provide it, but he remains

> hopelessly in danger of those that lurk in the night if not for the dog and his keen sense of smell and hearing to sound an alarm. Neither exists well alone. I could have called this painting 'The Archers' because both Leben and Tom were born under the zodiac fire sign of Sagittarius, nicknamed the Archer – always pointing his arrows towards a higher authority of which fire is one.[2]

These tender and touching observations move from direct observation to reverie and finally to myth. In their candour, they present a vision of fire that is the very antithesis of horror and devastation. One that is summed up in a single word, Adams's word – 'enchantment'; a term that well describes Nick Moir's photograph on the front page of the *Sydney Morning Herald*, illustrating a story of the fires that swept New South Wales, Australia, in the first days of 2006 (illus. 43). It shows two residents of Umina Heights gazing at a blazing landscape; if their

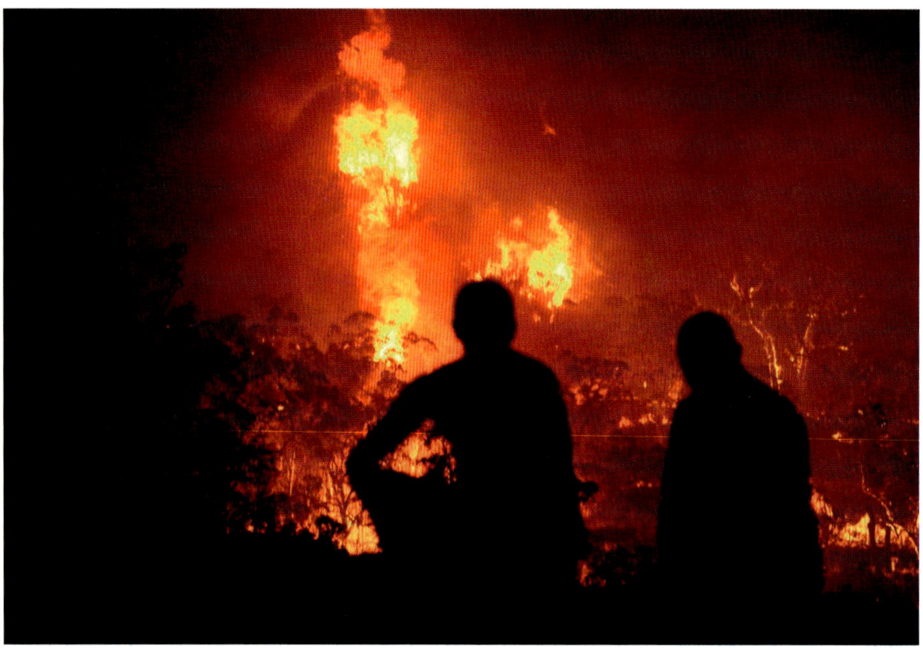

43 Nick Moir, photograph of fires at Umina, *Sydney Morning Herald*, 2 January 2006.

body language is anything to go by, they seem untroubled, spellbound even. The photograph, however, contrasts with their alarming circumstances, which, as the *Sydney Morning Herald* reported, saw residents battling throughout the night 'to keep showers of embers from hitting their homes'.[3] Moir's image, of course, has little to do with the subject of Suzanne Adams's painting: a controlled, small bonfire, by which a man and his dog stand. Yet, both works locate fire in the context of intimacy: through the shared personal experience of its intensity, on the one hand, and the visual and physical delights that it can offer, on the other. And both speak of fire's capacity to enthral and raze.

Hearth and Home: Vitruvius and Alberti

Luis Fernández-Galiano has cited in his recent book *Fire and Memory: On Architecture and Energy* (2000) a parable attributed to the distinguished architectural historian Reyner Banham, who recounts the tale of a 'primitive' tribe faced with a dilemma: what to do with 'fallen branches and some wood' – construct a shelter or use the materials to make a bonfire?[4] Fernández-Galiano suggests that

> both strategies are feasible. Both present themselves in most cultures and both deserve to be called architecture. Indeed, the thermal space of the bonfire is no less architectural than the visual space of the hut. Only an obstinate fetishism for icons or an object-oriented, hieratic conception of architecture can deny the bonfire the status of *ab ovo* [from the beginning] architecture so easily assigned to the hut. What is a house but a hearth?[5]

'What is a house but a hearth?' – a question that had been addressed many hundreds of years earlier by the renowned Augustan architect Vitruvius (*fl.* later 1st century BC) and, following him, Leon Battista Alberti (1404–1472). Vitruvius' *De Architectura libri decem* (*Ten Books on Architecture*) moves from considerations of building materials to public and private buildings and, finally, to water, sundials, clocks and machines. Early in Book Two, it addresses the discovery of fire:

> Humans, by their most ancient custom, were born like beasts in the woods, and caves, and groves . . . During this time . . . dense, close-growing trees stirred by stormy woods and rubbing their branches against one another, took fire. Terrified by the flames, those who were in the vicinity fled. Later, however, approaching more closely, when they discovered that the heat of the fire was a great advantage to the body, they threw logs into it and preserving it by this means they summoned others, showing what benefits they had from this thing by means of gestures. In this gathering of people, as they poured forth their breath in varying voices, they established words by happening upon them in their daily routine. Later, by signifying things with more frequent practice, they began by chance occurrence to speak sentences and thus produced conversations amongst themselves.[6]

Vitruvius' claims are very clear: the chance rubbing of branches during a storm led to fire which, while terrifying to those who first encountered it, soon heated their bodies so agreeably that 'this gathering of people' were given a voice; literally, to articulate the wonders of the new phenomenon. In this equation, the discovery of fire and the commencement of language go hand in hand. And it is from this heady mix that the 'building' emerges:

> The beginnings of association among human beings, their meeting and living together, thus came into being because of the discovery of fire. When many people came into a single place, having, beyond

> all the other animals, this gift of nature . . . Some in the group began to make coverings of leaves, others to dig caves under the mountains . . . Then, observing each other's homes and adding new ideas to their own, they created better types of houses as the days went by.[7]

Vitruvius' views on the origins of fire and its relationship to architecture had precedents in the earlier ideas of the Greek Hephaestus and Homer, but it is Vitruvius who first brings these notions together in a sustained thesis on the art of building. Some fifteen centuries later, Alberti would write *De re aedificatoria* (*On the Art of Building in Ten Books*), the first book on architecture since antiquity. On fire, he at first appears equivocal. He begins by saying that '"a roof and walls" mark the beginning of the congregation of men, and not, as "some have said," water or fire'. But a few pages later, he opines:

> In the beginning, men sought a place to rest in some region safe from danger; having found a place both suitable and agreeable, they took possession of the site. Not wishing to have all their household and private affairs conducted in the same place, they set aside one space for sleeping, another for the hearth, and allocated other spaces to different uses. After this men began to consider how to build a roof, as a shelter from the sun and the rain. For this purpose they built walls on which a roof could be laid.[8]

Alberti's hearth comes before the building of roofs and walls, that much is evident; but it is simply one of two other spaces fulfilling this role: the others as he says are 'for sleeping' and, in more generalized terms, 'other spaces for allocated use'. Elsewhere in his treatise, however, Alberti acknowledges the importance of the hearth emphatically:

> If we are to build somewhere very cold we shall need fires. There are different kinds of fires, but the most convenient is the bright, open hearth . . . What is more, the very sight of the light and flames of a live fire is a cheerful companion, they say, to old men chatting by the hearth.[9]

In this final observation, presented almost as an afterthought, Alberti removes fire from its utilitarian functions and relocates it in an imaginative order, one that Gaston Bachelard would much later describe in terms of 'reverie'. And one that he typically couches in terms of fire's ability to do one thing and yet suggest the other:

> The fire confined to the fireplace was no doubt the first object of reverie . . . One can hardly conceive of a philosophy of fire that would not include a reverie before a flaming log fire . . . To be sure, a fire warms us and gives us comfort. But one only becomes fully aware of this comforting sensation after quite a long period of contemplation of the flames . . . Of course the supporters of the theory of the utilitarian formation of the mind will not accept a theory so facile in its idealism, and they will point out to us the multiple uses of fire in order to ascertain the exact interest that we have in it: not only does fire give heat, but it also cooks meat! As if the complex hearth, peasant's hearth, precluded reverie![10]

So what then to make of the hearth? Historians would agree 'that some hearths were rich and others poor: beggars had hearths, the *menus* (little people) had hearths, as did the *gros* (the "fat cats", so to speak)'.[11] Hearths, in other words, were democratic in origin. Fireplaces set in walls, meanwhile, first appeared in the twelfth century. Before then, the round hearth was located in the centre of a room, and was used only for cooking. In other words, fire and its enjoyment was related strongly to its utilitarian outcomes.[12] In affluent households, however, fireplaces

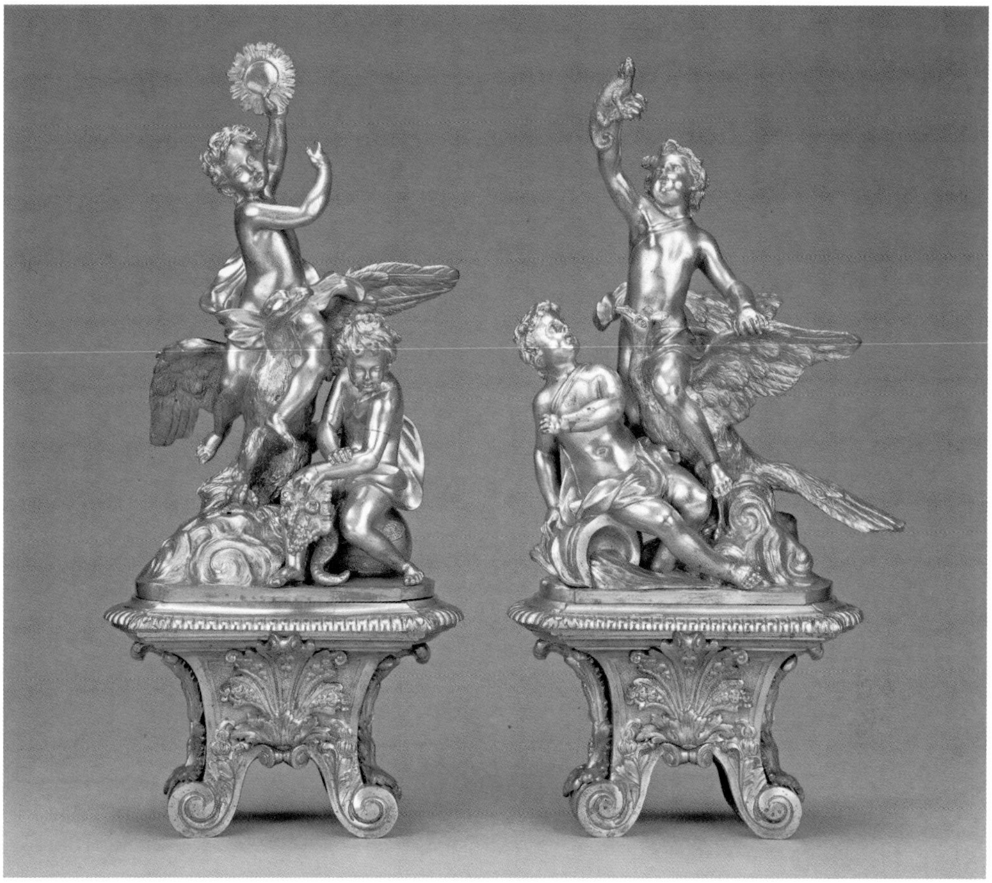

44 A pair of French firedogs, 1690–1715, gilt bronze.

would soon be transformed into decorative features.[13] Take this pair of *chenets* (the name derives from the French word for a small dog) from about 1700 (illus. 44).[14] Designed to hold logs, the ones illustrated here were intended simply to be ornamental. They support four children positioned with the attributes of the elements: the sun for Fire, a cornucopia and an orb for Earth, a vessel for Water and a chameleon for Air. Additionally, there is an eagle symbolizing Jupiter and a peacock representing his spouse, Juno. At work here is an iconography that points to an audience well versed in classical symbolism. The history of the domesticated fire thus embraces the twin (and enduring) necessities of heating and cooking, but it is also a history that shows how the functional can be transformed into the rarefied. The latter, of course, is a far remove from the simple fire that facilitates cooking or warming and the musings that both may conjure up.

Small Fires but Big Meanings: Some Dutch 'Little Masters'

Small fires may take the form of the hearth that is depicted in the miniature from the Arenberg Hours discussed at the beginning of this book (see illus. 1): here the man warming himself in front of the fire enlivens it, whereas in *Hearth* (1957), a large painting by the American Loren MacIver (1909–1998), the human presence is implied (illus. 45). Dominating this picture is the metal grate: holding an orange-red fire, it is set against a stuccoed wall covered in places by the blackening stains of soot. The painting is all crackles and pops and hot sights and textured surfaces. Standing alongside other viewers in front of MacIver's work (in the Metropolitan Museum of Art), I overheard two elderly visitors speaking: 'You can almost feel the heat', said one. The other mentioned 'marshmallows'. (Developing into a commercial product in the late nineteenth century, these tiny edibles are closely connected with the campsite and the hearth, at least in European and North American cultures. Soft and yielding, they complement well the gentle and ever-shifting nature of these little fires). Fastforward some 40 years from MacIver's *Hearth* to the Australian Wendy Sharpe (*b.* 1960) who paints herself comfortably ensconced in a chair and gazing at a fireplace (illus. 46). Executed while she was on a residency at the Haefliger Cottage in Hill End, New South Wales (owned by the artists Jean Bellette and her husband, Paul Haefliger in the 1940s), it's a robustly painted work that belies the quietly reflective nature of its subject.

This fascination with the domestic fire is nowhere more evident than in the seventeenth century, when Dutch painters turned time and again to small, interior scenes lit by candlelight, oil lamps, foot-warmers and fireplaces.

Consider Willem Duyster's *Soldiers Around a Fire* (illus. 47), where a figure with a wide-brimmed hat raises one hand to a candle held in the other.[15] Seeping through his fingers, the light illuminates part of his handsome breastplate. Looking to our right, his expression may best be

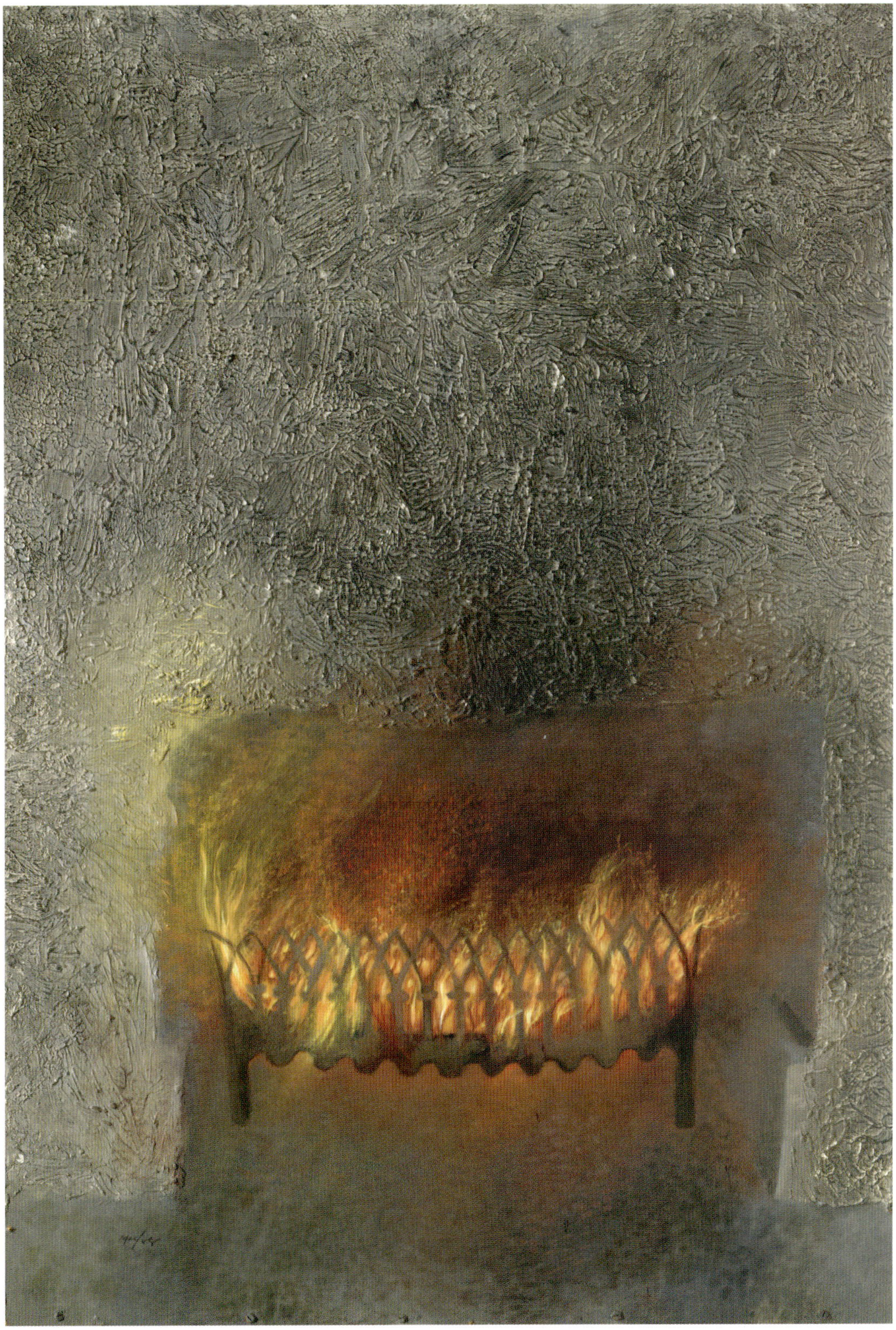

45 Loren MacIver, *Hearth*, 1957, oil and plaster on Masonite.

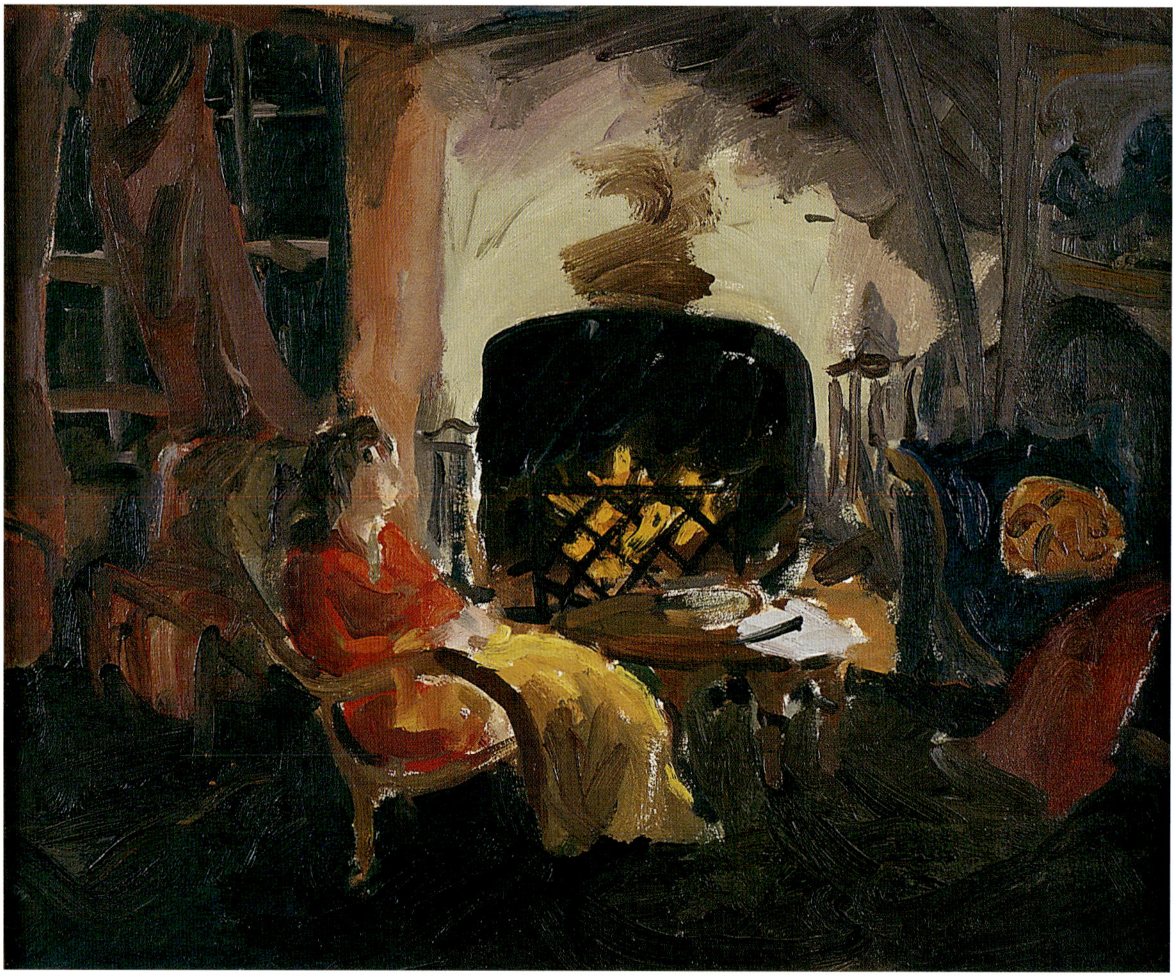

46 Wendy Sharpe, *Hill End*, 1994, oil on canvas.

described as one of melancholic absentmindedness. It throws into contrast, literally, the shadowed figure of a kneeling soldier who, not immediately seen, stares intently at a hearth; he stokes it and embers fly.

If Duyster's central figure is lost in his thoughts, then the aspiring artist in Gerrit Dou's *Young Man Drawing after a Sculpture* (illus. 48) appears to be mesmerized by the object of his attention, a plaster sculpture of a putto. Both are depicted in deep shadow, and parts of their respective anatomies catch the soft light of an oil lamp: the face and right hand of

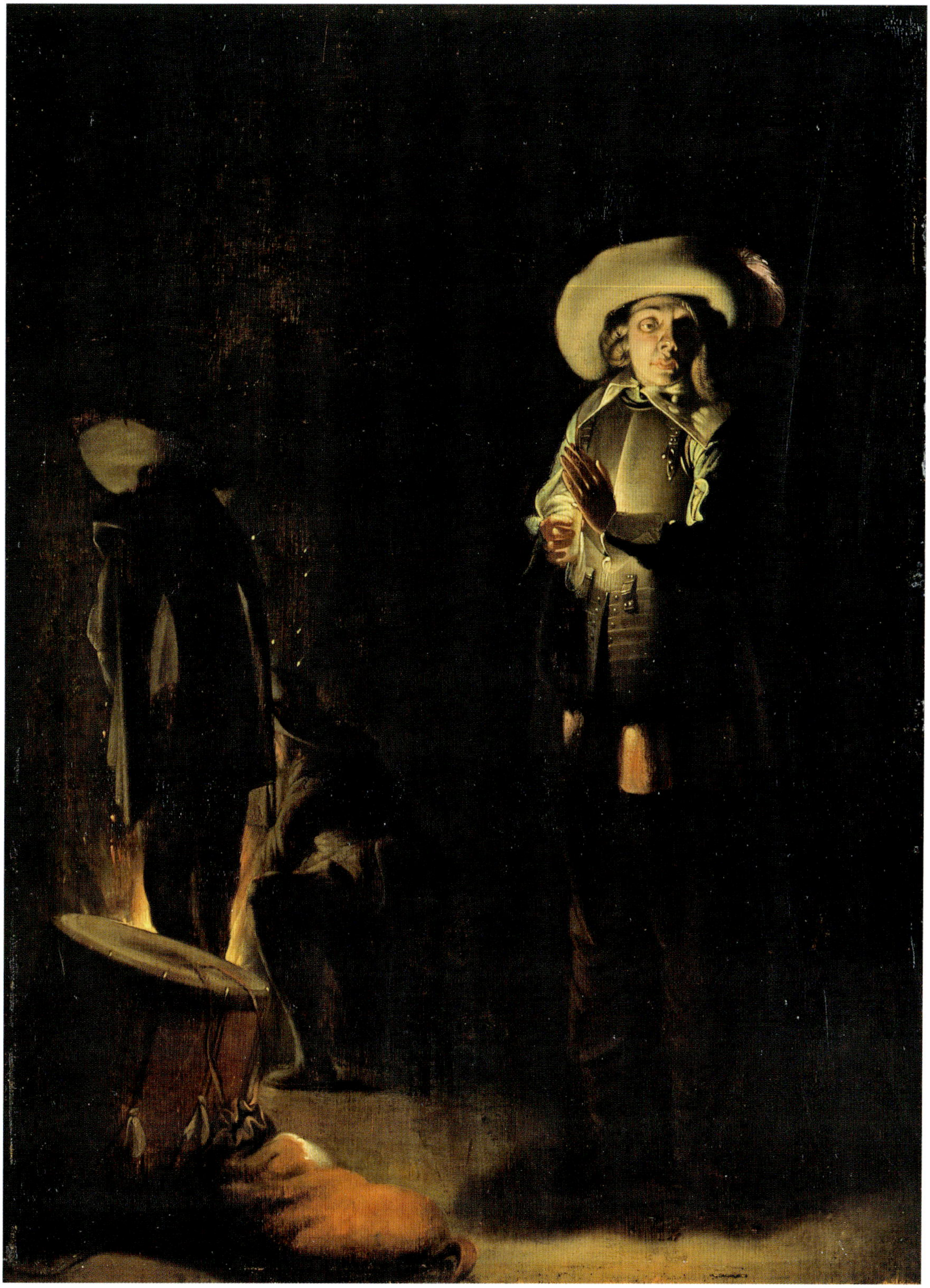

47 Willem Cornelisz. Duyster, *Soldiers Around a Fire*, *c.* 1630, oil on panel.

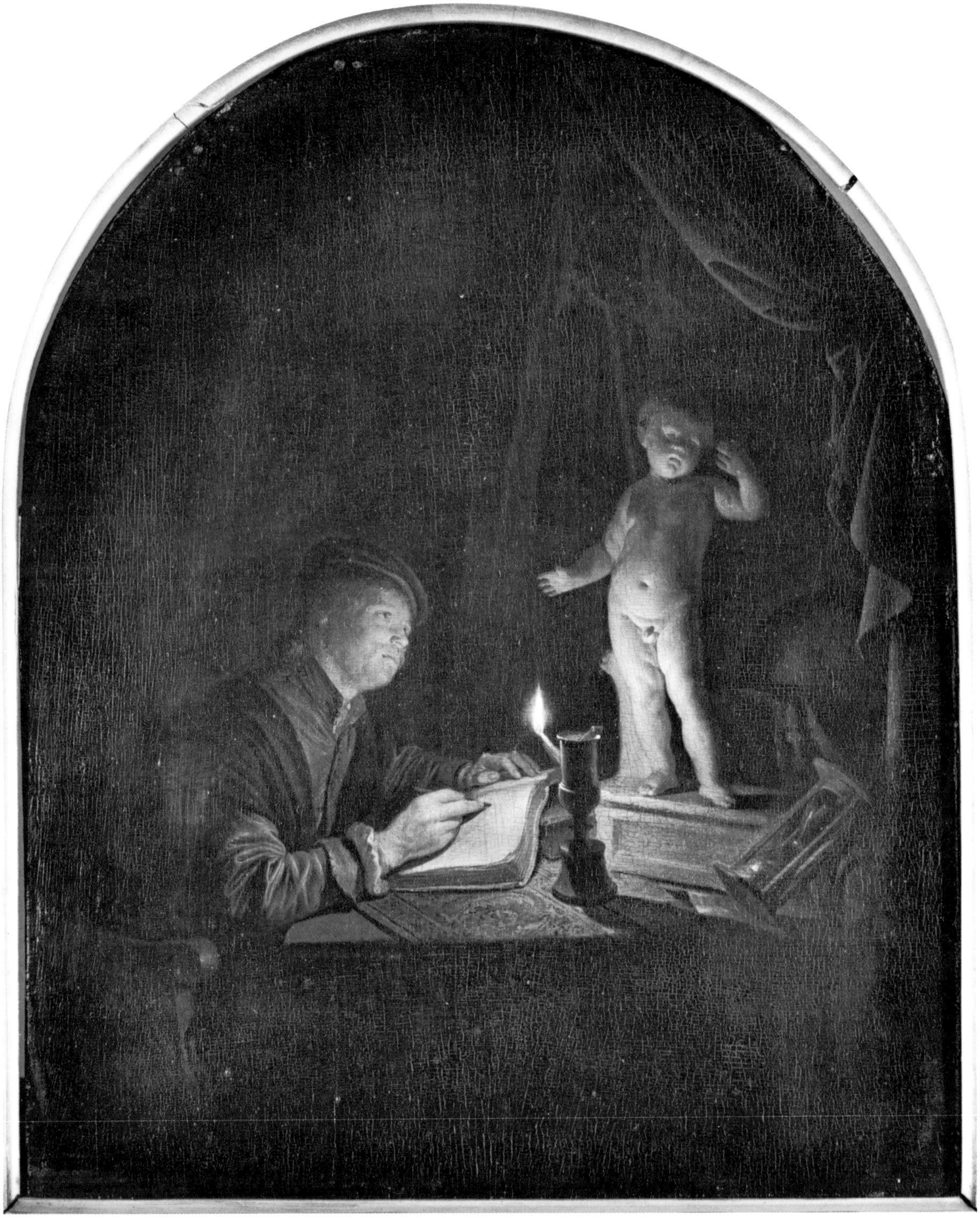

48 Gerrit Dou, *Young Man Drawing after a Sculpture*, 1650–55, oil on oak panel.

the student (including his sketch pad) and most of the putto's body. Here fire, on one level, is simply something that throws light, as it does in Dou's comparable *The Night School* (illus. 49), where there is a lantern on the floor, a candle held by the girl on the left, and the faint glow of another candle that falls over ill-defined figures in the background. A boy works on a slate, a young woman reads and the teacher hovers above them. Yet, as Friso Lammertse has importantly noted: 'Pictures by Dou and others of artists working by the light of a candle or an oil lamp [may also be] interpreted as metaphors of the concept "exercise" or "study".'[16] Lammertse continues:

> According to Cesare Ripa [a seventeenth-century writer], 'Exercise in the arts, study' was to be depicted by 'A pale-faced boy, modestly dressed and seated. With his left hand, he shall open a book, which he reads with concentration, having in his right hand a pen, and at his side a lit lamp.'[17]

Far removed from these two paintings of 'study' aided by fire is Godfried Schalcken's *Boy Blowing on a Firebrand to Light a Candle*, illus. 50). A pupil of Samuel van Hoogstraten (1643–1706) and then of Dou in Leiden, Schalcken shows an adolescent, lips pursed, blowing on a glowing stick; he holds an unlit candle in his right hand which appears strangely enlarged. Intensely focused on his activity, his face is lit by the flame that flies off the firebrand. The painting is in fact a recreation of a classical subject painted by Antipholos of Alexandria that is described by Pliny in his *Natural History*.[18] A motif popular among many of Schalcken's contemporaries, it had also been depicted by the Cretan El Greco (*c.* 1541–1614) some 100 years earlier. In his *An Allegory (Fábula)*, El Greco shows an androgynous figure lighting a small candle that is held up against a glowing ember (illus. 51). Joined by a monkey and a man in a yellow garment and red hat, figures whose presence continue to invite all sorts of interpretation, our eyes are directed to the central

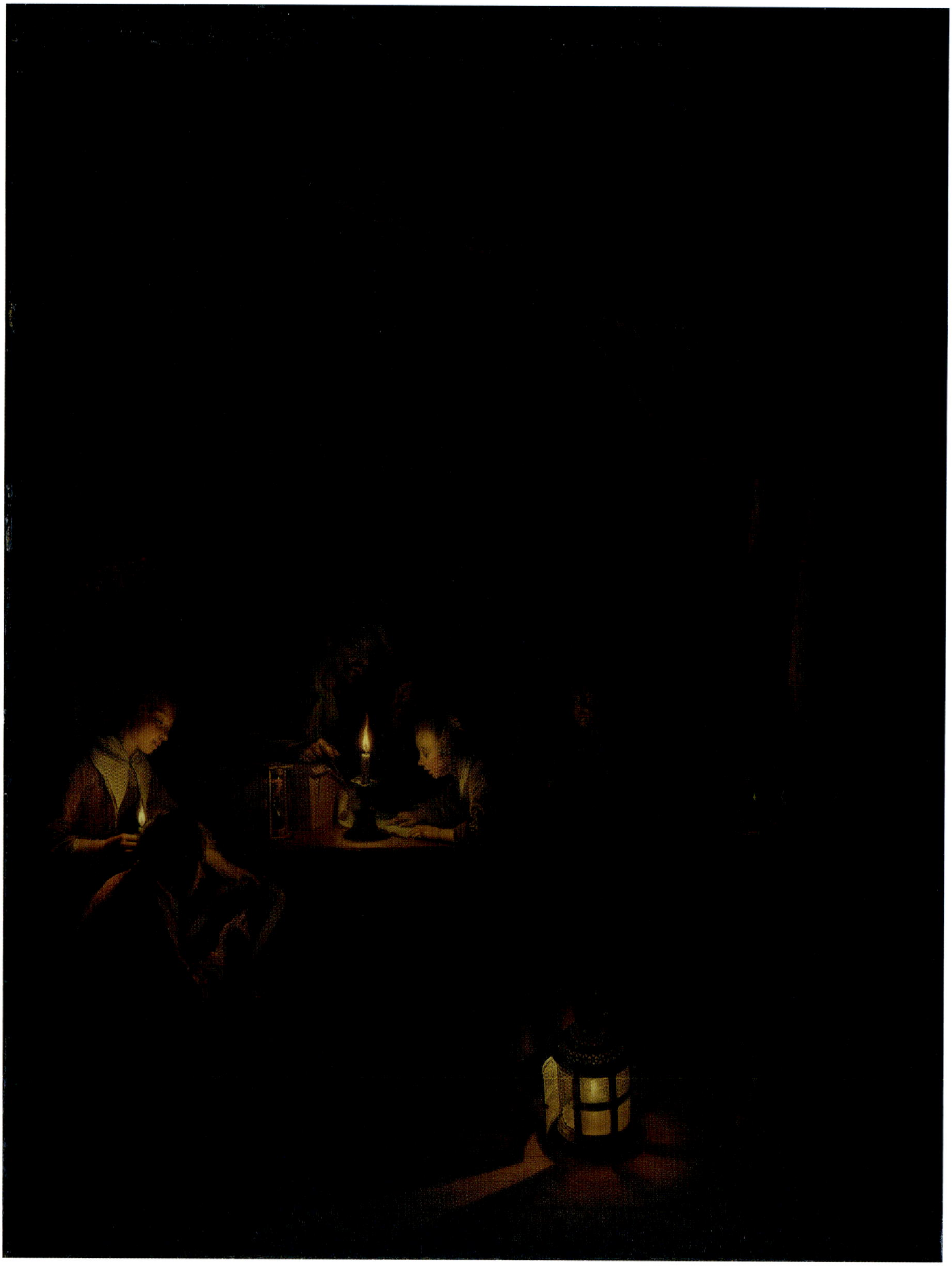

49 Gerrit Dou, *The Night School*, before 1665, oil on panel.

act of ignition, as it were.[19] This painting, it seems to me, takes us back to Prometheus, to the 'giving of fire' to humankind and to the wisdom that it bestows: there is the star-struck monkey, the goggle-eyed man (a 'fool', as he has sometimes been described) and the androgynous figure who quietly and diligently goes about the task of creating light. Bringing together animal and simpleton, male and female, all transfixed by the

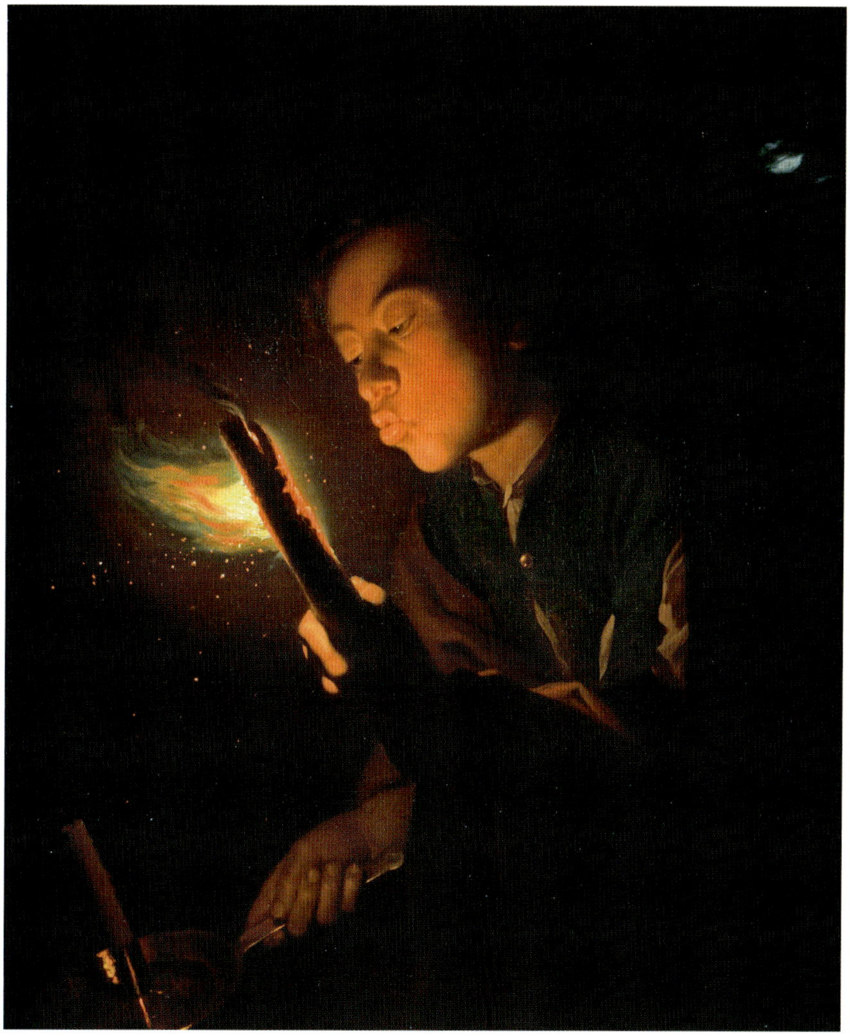

50 Godfried Schalcken, *Boy Blowing on a Firebrand to Light a Candle*, c. 1692–8, oil on canvas.

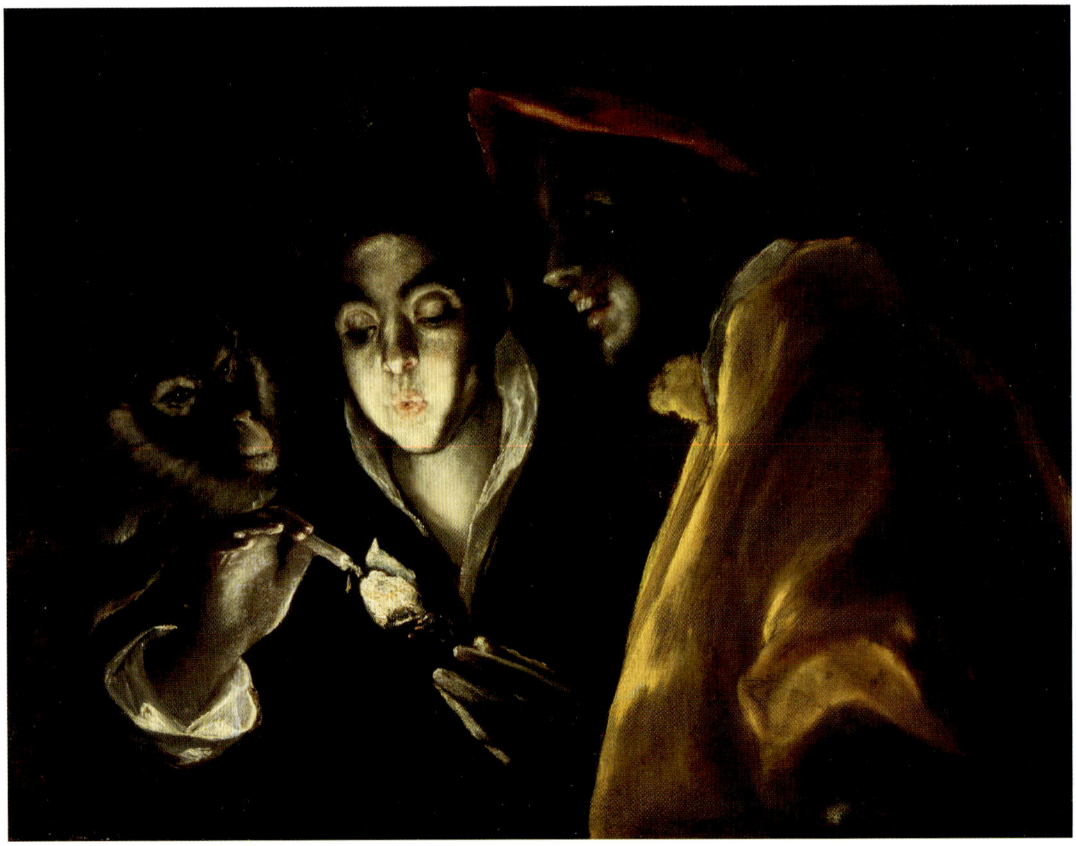

51 El Greco, *An Allegory (Fábula)*, *c.* 1580–85, oil on canvas.

act of making fire, El Greco's *An Allegory* combines the wondrous nature of this element with its potential to (literally) enlighten.

Returning to the Dutch 'little masters', there is Judith Leyster's *The Proposition* (illus. 52). Leyster was born in Haarlem in 1609. Here she depicts a man leaning over a modestly dressed woman engrossed in sewing. Light from an oil lamp on the table illuminates their faces and throws shadows on the wall. A foot-warmer with its coals glowing is visible under her skirt. The man, who wears an 'un-dutch fur hat, indicating that he is a foreigner',[20] holds out his right hand containing some coins while the other, palm down, gently but rather menacingly touches the

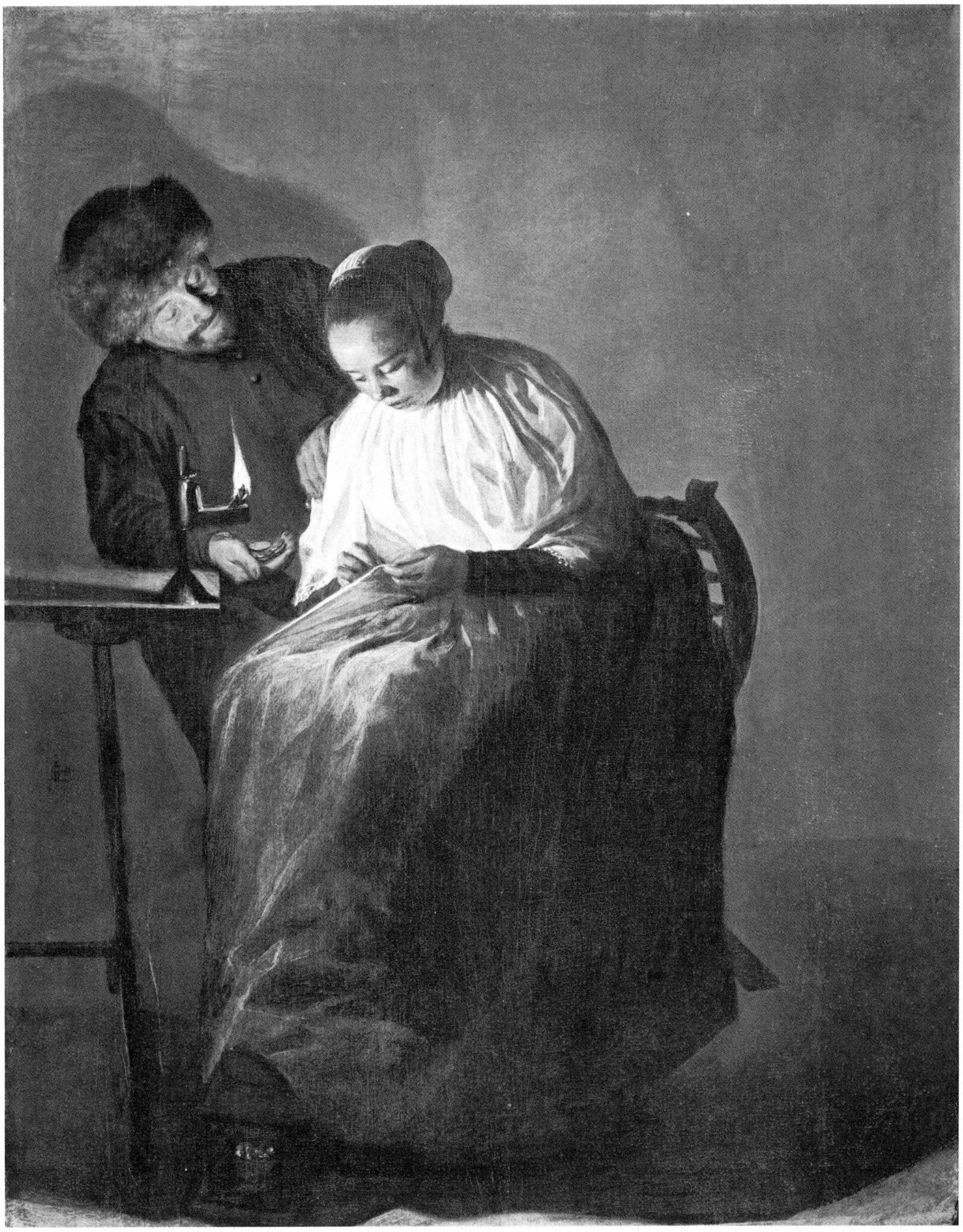

52 Judith Leyster, *The Proposition*, 1631, oil on panel.

woman's shoulder. He has a lecherous look; she, impassive, goes about her activity unwaveringly. This depiction of fire has no relation to the symbolic purposes to which it is put in the work of Dou and Schalcken, discussed previously, where it served a pedagogic role. Rather, in *The Proposition*, it helps frame the context for sexual manoeuvrings. The modest foot-warmer is of particular interest. As the art historian Ann Sutherland Harris has observed:

> When shown under the skirt [as it is here] the lady is taken; halfway out with her foot on the warmer, she may be signalling that a suitor has a chance. If it is on the floor and empty [as in Vermeer's well-known *The Milkmaid*, c. 1658–60] she currently has no male companion.[21]

Striking Matches: Fact and Fancy

The striking of a match did not produce the tiny flames that are depicted in seventeenth-century painting, or, for that matter, in any of the fires, great or small, private or public, that we may come across anywhere before the early decades of the nineteenth century. This singular experience would require the isolation of phosphorus in 1669, and then, *inter alia*, the sharp-witted inventions, respectively, of an Irish physicist named Robert Boyles (1680), and then, some 150 years later, of an English chemist, John Walker (1827), and Samuel Jones of the Strand, London (1828), who was the first to recognize the marketable potential of matches.[22]

John French interrupts this chronology with his extraordinary *The Art of Alchemy*, published in London in 1651. Tantalizingly, he anticipates matches by describing how fire could be contained in a glass that 'while shut will not burn, but as soon as it is opened will be inflamed'.[23] Involving, amongst other procedures, the extraction of 'the burning spirit of the salt of tin' in a retort, this amazing proposal, French opined, 'may be of many uses . . . for any man may carry it about with him and let it burn on a sudden when he has any occasion for fire'.[24]

This wonderfully abstruse concept invites comparison with the 'burning glasses' developed by early alchemists and chemists. Consisting of lenses positioned in the sun and angled towards 'the substance to be heated, a heat source was created which avoided the need to place an object directly in a fire'.[25] A bookplate engraved by J. Pass at the end of the eighteenth century illustrates some of these remarkable devices (illus. 53). The ones he depicts speak of an Enlightenment culture of clinical observation where sharply delineated images recall scientific instruments. With one exception, the small, rather pathetic smouldering emanating from a nondescript blob, it's as if the very product of these 'burning glasses', fire itself, was still to be regarded as something verging on what we may now describe as the abject. In contrast, there is the contemporaneous, typically domestic English tinder box, consisting of a circular box and candle socket on the end where the flint would be struck against the metal to create sparks and ignite the dry material (illus. 54). Straightforward in design, the device is far removed from the more elaborate European tinder pistol (illus. 55). Looking to all intents and purposes like a (mechanical) grasshopper, the name by which these mechanisms are now called, it was created for affluent households.[26] Inspired by ordinary pistols that were then often converted by a gunsmith, the 'grasshopper' and other similar contraptions transformed a piece of weaponry into a rarefied object of desire, in ways that are redolent of the French firedogs or *chenets* (*c*. 1700), discussed earlier in this chapter. And so it is with this Japanese fire-making device, a clever adaptation of a *netsuke* (the clasp used to attach a small box to the sash of a kimono), in which a tinder pistol is ensconced in an egg-shaped case (illus. 56).[27]

Guns and clothing, the troublesome and the comforting – fire's inherent polarities – return us to Robert Boyles, the Irish physicist named

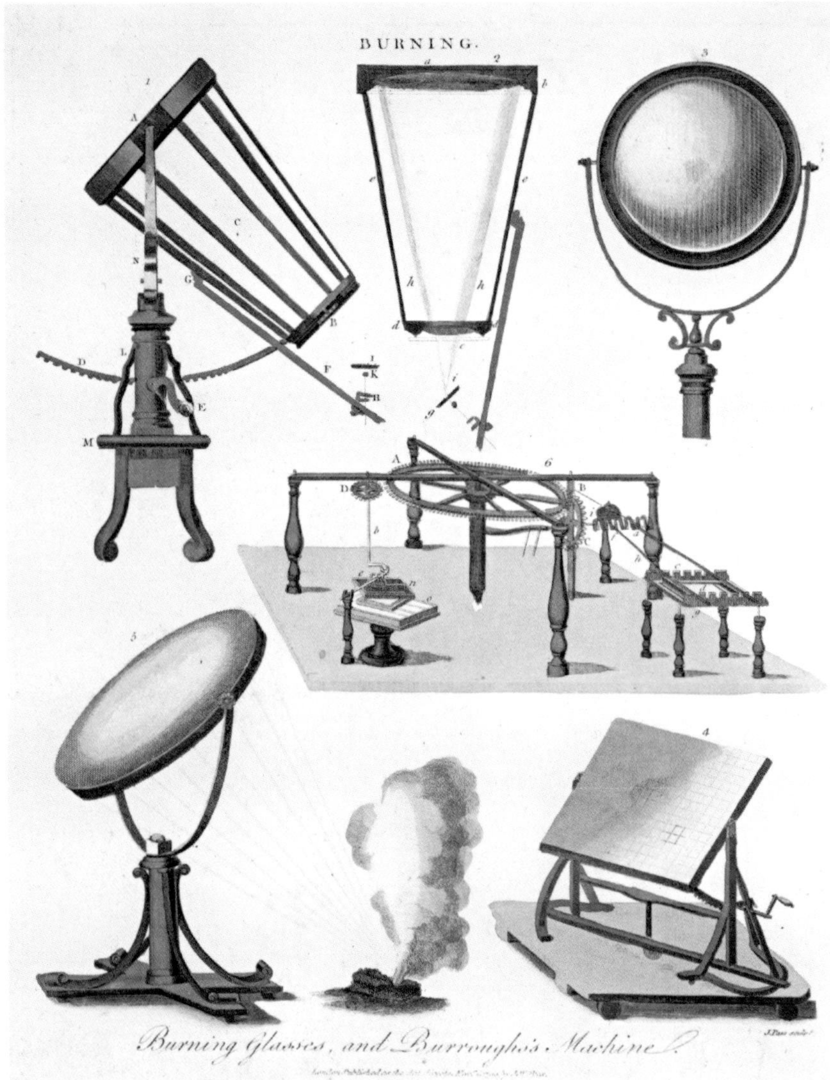

53 Bookplate illustrating 'Burning Glasses' and 'Burroughs's Machine', 1799.

above, who discovered that he could create fire by coating a piece of wood with sulphur and rubbing it over a piece of paper covered with phosphorus. It was left to John Walker, however, to come up with an actual match: wood splinters dipped in sulphur and tipped with a combination of antimony sulphide, potassium chlorate, gum and starch, which, when rubbed against an abrasive surface, produced a flame.[28] Surprisingly, he did not patent his invention; but he did have the gumption to designate it 'Congreve's', after the rocket of the same name. Conceived as an offensive tool by a British ordnance officer, William Congreve, the rocket was first employed in 1806 during the Napoleonic

54 An English tinder box, *c.* 1750–1820.

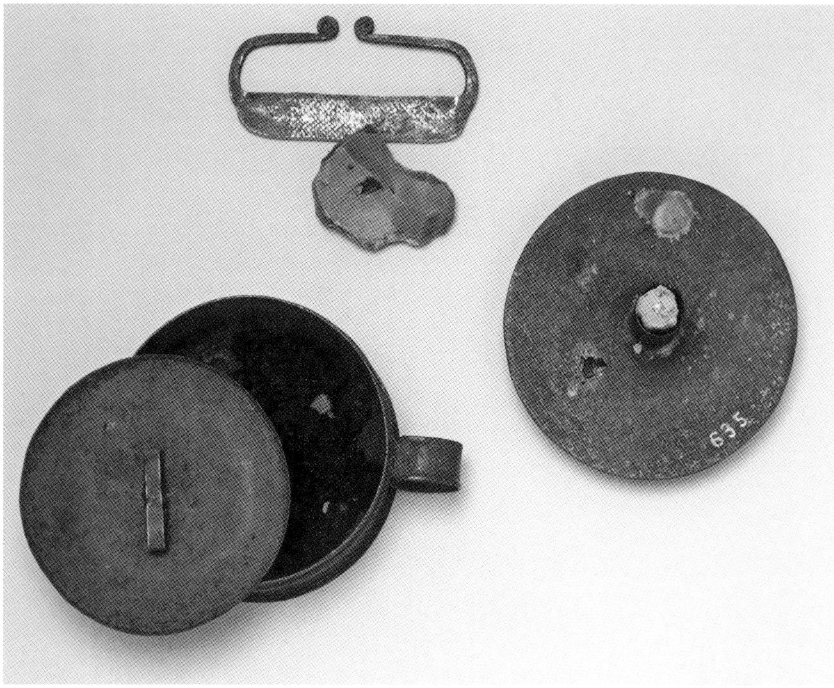

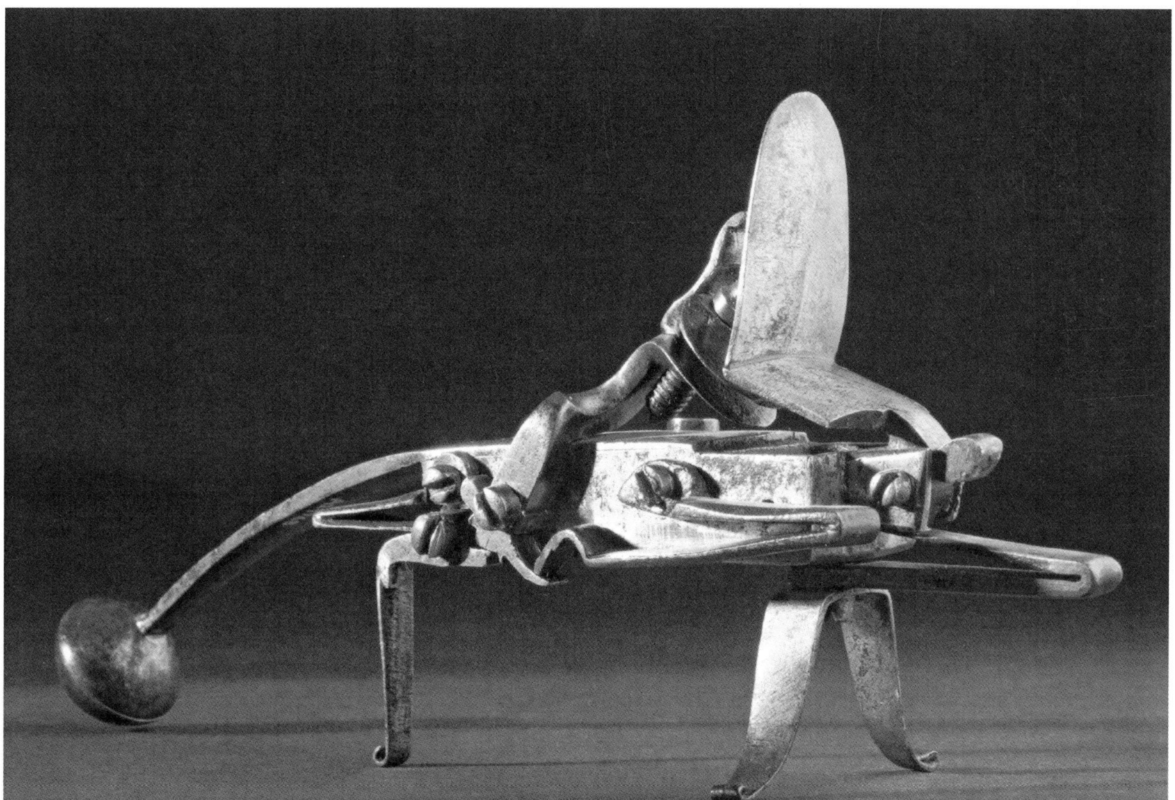

55 An all-steel European grasshopper tinder pistol.

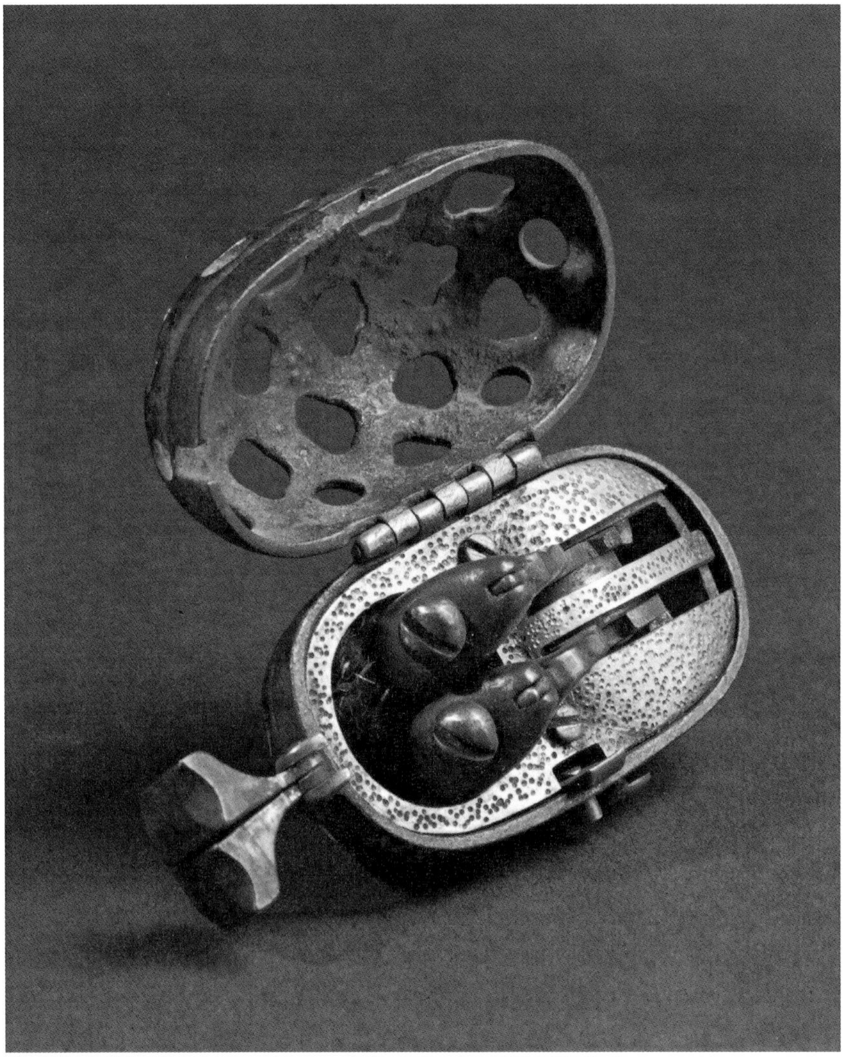

56 A Japanese tinder pistol in egg-shaped case.

War.[29] If Walker located his 'friction matches' symbolically in the sphere of contemporary weaponry (as indeed were all the tinder pistols described above), then Samuel Jones moved tantalizingly between other imaginative orders. First, there were his 'Prometheans'. Patented in 1828, they consisted of individual glass sacs containing a drop of vitriol, which were sealed and then wrapped in paper coated with chlorate of potash, sugar and gum arabic. Snapping the match with a pair of pliers broke the small glass container and allowed the vitriol to mix with the primed paper thus creating a flame.[30] Jones followed these with his 'Lucifers', identical to Walker's, except for their designation that, of course, was indebted to the name of the fallen angel of medieval times associated with Satan. Yet even here there is a slight ambiguity, since the word 'Lucifer' is derived both from the Latin term, 'light-bearer', and the Greek *heosphorus*, 'dawn bearer'. Christian interpreters, however, believe that the 'son of the morning' (from Isaiah 14:12) is a reference to Satan.[31]

The early history of matchmaking is enmeshed in what is the recurring theme of this book: fire's contradictory effects and meanings. The debilitating consequences of these tiny producers of fire, however, left little to the imagination. The French chemist Charles Sauria made a match with white phosphorus in 1830, which resulted in toothache and the swelling of the gums and jaw, dubbed 'phossy jaw' (narcosis of the jaw). On the box containing Jones's 'Lucifers', there was this warning: 'If possible, avoid inhaling gas that escapes from the combustion of the black composition. Persons whose lungs are delicate should by no means use lucifers.'[32] The widespread use of yellow phosphorus was similarly toxic, alarming the feminist Catherine Booth (née Brixton), who was equally concerned about the very low wages paid to female matchmakers in London (one shilling and four pence for a sixteen-hour day). A year after her death in 1890, her husband, William Booth, the founder of the Salvation Army, set up a match factory that would replace yellow with red phosphorus. Booth called his matches, fittingly, 'Lights in Darkest England'.[33]

The production and marketing of matches impacted on popular culture. A telling example is the children's tale '*Dir gar traurige Geschichte mit dem Feuerzeug*' ('The Dreadful Story of Harriet and the Matches'), written by the Frankfurt physician Heinrich Hoffmann (1809–1894). Wishing to purchase a Christmas gift of a book for his three-year-old son, but unimpressed by those that were available, he instead created his own, the first edition of which was published in 1844 as *Lustige Geschichten und drollige Bilder* [*Merry stories and funny pictures*] *mit 15 schön kolorierten Tafeln für Kinder von 3 bis 6 Jahren.* In addition to the story of Harriet and her matches, the collection, now generally known as *Struwwelpeter*, included such tempting titles as 'Cruel Frederick', 'The Story of Little Suck-a-Thumb' and 'The Story of Johnny Head-in-Air'.

Hoffmann's story of Harriet translates fire's contradictory effects into a tale of innocence, desire and death that brings to mind the fairy tales of the Brothers Grimm, where the grotesque and the jocular sit side by side.[34]

Hoffmann takes an innocent girl and places her in a combustible situation (illus. 57):

> It almost makes me cry to tell
> What foolish Harriet befell.
> Mamma and Nurse went out one day,
> And left her all alone at play.
> Now on the table close at hand,
> A box of matches chanced to stand;
> And kind Mamma and Nurse had told her,
> That, if she touched them, they would scold her.
> But Harriet said, 'Oh, what a pity!
> For when they burn, it is so pretty;
> They crackle so, and spit, and flame:
> Mamma, too, often does the same.'
>
> The pussy-cats heard this,
> And they began to hiss,
> And stretch their claws,
> And raise their paws;

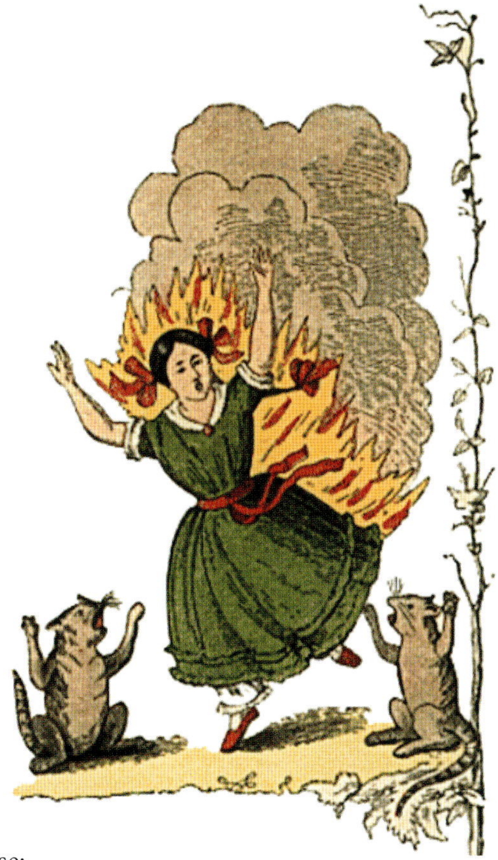

57 'The Dreadful Story of Harriet and the Matches', from Heinrich Hoffmann's *Struwwelpeter*.

'Me-ow,' they said, 'me-ow, me-o,
You'll burn to death, if you do so.'

But Harriet would not take advice:
She lit a match, it was so nice!
It crackled so, it burned so clear –
Exactly like the picture here . . .

And see! oh what a dreadful thing!
The fire has caught her apron-string;
Her apron burns, her arms, her hair –
She burns all over, everywhere . . .

So she was burnt, with all her clothes,
And arms and hands, and eyes and nose;
Till she had nothing more to lose
Except her little scarlet shoes;
And nothing else but these was found
Among her ashes on the ground.

And when the good cats sat beside
The smoking ashes, how they cried![35]

This tale brings together a number of more obvious themes: parental dictate, a child's capriciousness and matches. Combined, their outcome is 'dreadful': the horrendous demise of Harriet, burnt tooth and nail so that only her 'smoking ashes' remain. The story could end there, of course, but our sensitivities today to the rich field of children's texts (written, needless to say, by adults) invite other, more elaborate, readings. Harriet's first encounter with fire is all about aesthetic delight ('it is so pretty'). It is also one coupled with a fierce independence: she had turned her back on the warnings of her parents and nurse. Then enter the two cats, which similarly urge her to desist. Even they, however, cannot dissuade Harriet; and the result, as we know, is disastrous, illustrated in the final image showing the two felines shedding tears over the fantastic re-presentation of Harriet's burnt body. Fire and its inescapable associations with allure and ruin, nature and culture, are the sub-texts of this tale.

Some two decades after the publication of 'Harriet and the Matches', reality came disturbingly close to fantasy, when Archduchess Matilda of Austria trod accidentally on a match and was burnt alive. London's *The Times* newspaper would pick up on Matilda's misadventure and describe it in phlegmatic ways:

> The archduchess Matilda has ceased from suffering. The intended Mother of the future Kings of Italy . . . the daughter of the conqueror of Custozza, and grand-daughter of the hero of Aspern and Essling, a Princess in her nineteenth year . . . died on Thursday last at 8 o'clock in the morning of a Lucifer match. She inadvertently trod on one which was lying at her feet on the floor as she leant out at the window talking to one of her relatives; her summer dress was in a blaze before she was aware of it, and before anyone could run to her rescue she sank to the ground in agony of pain from which only death released her.[36]

The fact that Archduchess Matilda met her fate as a result of a 'Lucifer' match played into the hands of the burgeoning matchmaking business of William Bryant and Francis May, which they set up in London's East End in 1861.[37] Granted a UK patent in 1855 to produce a safety match developed by the Lundström brothers of Sweden, Bryant & May was quick to respond to Matilda's bad luck. They placed an advertisement in *The Tomahawk* that included the above extract from *The Times* (with those passages dealing with the 'Lucifer' and Matilda's fiery death suitably highlighted) plus their own disclaimer:

> The above ACCIDENT could not possibly have occurred with BRYANT AND MAY'S PATENT SAFETY MATCHES, WHICH LIGHT ONLY ON THE BOX in which they are contained, as they may be trodden upon or exposed to any ordinary degree of heat without becoming ignited.[38]

Between Hoffmann's story of little Harriet and Bryant & May's advertisement, the Danish children's writer, Hans Christian Andersen (1805–1875) penned his celebrated *The Little Match-Seller*. Published in 1846, this poignant story combines ruminations in front of fire with the contrasting themes of desire, celebration and death. It was a dark and very cold and snowy New Year's Eve, writes Andersen, when a 'poor little girl, with bare head and naked feet' came across a house in which 'lights were shining from every window'.[39] The smell of roast goose filled her senses. She decides to light a match to warm her chilled body:

> She drew one out – 'scratch'! how it spluttered as it burnt! It gave a warm, bright light, like a candle . . . It was really a wonderful light.

Here the visual and the physical are intimately balanced. The flames soon conjure up a vision of the girl sitting in front of an iron stove, all comfort and warmth. But just at this wondrous moment, the small flame goes out. Undeterred, she lights another match. This second striking is even more revealing: she is now transported to a 'table covered with a snow white table-cloth, on which stood a splendid dinner'. Yet again, the match is extinguished. So, she lights another, and now finds herself 'under a beautiful Christmas tree', where 'thousands of tapers were burning upon the green branches'. At this juncture, Andersen shifts his attention from the domestic and the private to the public, the 'outside', where the girl sees a falling star. In this instance she remembers her grandmother, now deceased, who had once told her that a falling star signals a soul going to heaven. Again, the girl strikes a match 'on the wall' (something that becomes, in Andersen's hands, a refrain in the tale). This time her grandmother appears before her, and, wishing to keep hold of this experience the girl strikes all her matches, which glow 'with a light that was brighter than the noon-day'. Then, arm-in-arm with her grandmother, they both 'flow upwards in brightness and joy'. The following morning the match-seller is found frozen to death, 'holding the matches in her hand, one bundle of which was not burnt'.

Fire Leaping

The Archduchess Matilda and Hoffmann's Harriet had no luck with matches, but there's a certain character called Jack whose encounter with tiny flames was far more agreeable: this intriguing figure appears in the short nursery rhyme, 'Jack Be Nimble', first published in 1798:

> Jack, be nimble,
> Jack, be quick,
> Jack, jump over
> The candlestick.[40]

The origins of this ditty are obscure and involve, variously, piracy, pagan rituals, religious beliefs and lace-making; sundry sources that are not unusual in nursery rhymes, but which in 'Jack Be Nimble' help tease out fire's often contradictory uses and meanings.

A 'Black Jack' is sometimes mentioned: he is (supposedly) a late sixteenth-century pirate well known for his ability to evade the authorities, hence 'Jack, be nimble'. Nothing in the literature on piracy, however, suggests that this pseudonymous figure ever existed, although there is a 'Blackbeard', the *nom de guerre* of the pirate Edward Teach, who died in 1718.[41] A pity, really, since the thought of an agile pirate metaphorically jumping over candlesticks (otherwise, 'the powers that be') is appealing.

Pagan rituals, on the other hand, might suggest more fruitful foundations. Chris Roberts has recently written about 'Jack Be Nimble' and its associations with fire jumping and fertility – an activity, he points out, popularized by *The Wicker Man*, the 1973 cult film directed by Robin Hardy, in which naked young women (they actually wore body stockings) leap 'over fire in order to promote their fecundity'.[42] Prancing maidens and female productivity, however, would seem to have little to do with the very gendered 'Jack' of the rhyme, but there is another long history of fire jumping in Britain, associated with good luck, that may help begin to contextualize Jack's origins. Take *Bealtine* (or *Bealltuinn*), the ancient Gaelic holiday. Describing this in his exhaustive Gaelic-English dictionary, first published between 1901 and 1911, Edward Dwelly says:

> May-day, first day of May. On the first of May was held a great Druidical Festival in favour of the god Belus. On this day fires were kindled on the mountain tops for the purposes of sacrifice; and between these fires the cattle were driven, to preserve them from contagion till next May-day. On this day it was usual to extinguish all the hearth fires, in order that they should be re-kindled from this purifying flame. In many parts of the Highlands the young folks of the district used to meet on the moors on 1st May. They cut a table in the green sod, of a round figure . . . They then kindled a fire, and dressed a repast of eggs and milk of the constituency of custard . . . After the custard was eaten, they divided the cake into as many portions as there were people. They daubed one of the pieces with charcoal . . . and [then the pieces] were all put into a bonnet . . . and each one [of the participants], blindfolded, drew out a portion . . . whoever drew the black bit was the devoted person who was to be sacrificed to Baal, whose favour they meant to implore in rendering the year productive. The devoted person was compelled to leap three times over the flames.[43]

These activities invite comparison with a similar practice in medieval England, when the celebration of weddings on 25 November, St Catherine's Day, is said to have concluded with the ritual jumping over a candlestick.[44] Extinguishing the flame would result in bad luck for the remainder of the year. This last explanation comes close to the scenario of 'Jack Be Nimble'. It also is one that segues into the practice of English lacemakers, which, on one of their most celebrated holidays (the same 'Catterns' or Catherine's day), involved boys and girls jumping over a lighted candlestick. (Catherine was the patron saint of spinners; but lacemakers also believed they were related to her).

According to Thomas Wright, the people of Wendover in Buckinghamshire

> called Catterns 'Candle Day', it being the first day on which they commenced to make lace by candle-light . . . At the lace schools the girls and boys danced in a ring round the great lace-maker's candlestick [over 60 cm/2 ft high, one of these is preserved in the Buckinghamshire County Museum, Aylesbury], singing:
>
> Wallflowers, wallflowers, growing up so high,
> All young maidens surely have to die;
> Excepting Emma Caudrey, she's the best of all.
> She can dance and she can skip,
> And she can jump the candlestick.[45]

Wright goes on to say that

> In other places, including West Suffolk, which was formerly a lace-making centre, the song was,
>
> > Jack, be nimble! Jack be quick!
> > Jack, jump over the candlestick . . .

> The name of any boy or girl was inserted in both the rhymes. The object it seems was to clear both the candlestick and candle without extinguishing the light. If the light was extinguished ill luck was supposed to follow during the subsequent twelve months.[46]

This merry game, it must be said, masks what is, after all, child labour. Working alongside adults in the lace-making counties of the East Midlands, children learnt their skills in the workhouses; from the commencement of the nineteenth century, however, some lace schools operated outside these institutions.[47] Channer and Roberts give some insight into the conditions that these young lacemakers toiled under in their 1900 study of the industry. While their chapter on lace schools is a self-confessed, imaginary account, and sentimentalized to boot, there is enough to suggest how dire circumstances were. We read about children between the ages of six and sixteen, 'sitting in rows, each little person on his or her four-legged stool, with its pillow resting against the three-legged stand in front of it'.[48] A teacher supervises them. 'To escape the cane they have to put in ten pins a minute, or 600 an hour.' Dinner-time arrives:

> Each child covers over its pillow . . . All but a girl of seven . . . She is faint and weak from one of those sick headaches to which poor children are so often subject . . . and, like all the others . . . has been working since six o'clock.

The account then moves to a 'winter's evening' at another school, where some sixteen girls are working by the 'light of one tallow candle'. The room, 'though stuffy, is cold, for, being so full, the fire has to be kept low'. We learn that the winter hours are usually from 'eight to eight, allowing two hours for meals, but many work an hour or so longer'. And what is the workers' pay? The teacher says that the 'girls at her school usually earn, after deducting what they pay her, *2s. 3d.* a week'. Anne Buck confirms these figures in her scholarly study of the East Midlands lace-making industry: taking, as an example, three lace schools in Chesham, Buckinghamshire, in 1819, she notes that children aged five could after two years earn between two shillings and two shillings and sixpence a week. Deductions from this amount included the extra charge for candles during the winter months.[49]

Ditties and games, good luck and bad – and child labour: all these coalesce around Jack and his exploits over the candlestick (at least in the Anglo-Saxon/Celtic origins of the rhyme). Yet there is another tradition that also makes much of fire jumping and for similar reasons: for Iranians, whose new year commences around the spring equinox, in late March, jumping over fires is a feature of the Chaharshanbeth-suri, which is held to 'banish the lingering bad luck of the dying year'.[50] Recalling this occasion growing up in Dezful, in southern Iran, Paliman Salimpour says:

> I am very competitive by nature, and the other girls were too foo-fooey, but I always wanted to compete with the boys. They would get naughty and add wood to the fire, and it would get bigger than it should be. I never told anyone, but I had really long hair then, and the tips of my hair got burned.[51]

But back to Jack and his candlestick. Like fire itself, the rhyme can move effortlessly into other contexts. Iona and Peter Opie, in their influential dictionary of nursery rhymes, draw attention to Ginette Dunn's research in which she describes how 'the candlestick dance' continued in Suffolk pubs in the early years of the last century.[52] Dunn quotes Bessie Ling (*b. c.* 1900), for example, who recalls doing 'the candlestick dance' at a pub called the Ship Inn in Blaxhall.[53] Ben Ling (*b.* 1891) remembers this dance also being performed in Blaxhall at Christmas, when they:

> Used to have a lighted candle on the floor and two of the old girls used to dance, pull their skirts up, you know, they got long skirts – pull their skirts up and do, 'Jack be nimble, Jack be quick, Jack be over the candlestick'. Yes, they used to enjoy themselves.[54]

Giving a flirtatious slant to the 'candlestick dance', Ben Ling's recollection hints at one of fire's great literary tropes – love and lust. 'Through fire everything changes',[55] writes Bachelard. It is this transformative nature of fire that lends itself easily to the machinations of sexual desire; and nowhere is this more apparent than in Charlotte Brontë's famous novel of 1847, *Jane Eyre.*

The Sexualized Fire: A Nineteenth-century Case Study

Presenting an unconventional view of a woman's role in the context of Victorian sensibilities, *Jane Eyre* traverses gender, nationality, class, religion and spirituality. Widely translated, it has been adapted by a host of filmmakers, including Theodore Marston (1910), Robert Stevenson (1943) and Franco Zeffirelli (1996). Television has had its fair share of adaptations, including those by Robert Young (1987) and Susannah White (2006). In addition, the English composer Michael Berkeley, with a libretto by the Australian writer David Malouf, turned the novel into a two-act opera (2000).[56]

In summary, *Jane Eyre* tells the story of the orphaned Jane who lives with her aunt, Sara Reed, at Gateshead. These are unhappy times: Jane receives little sympathy from her cousins, Georgina and Eliza, whose brother, John, is more openly antagonistic to her; the household physician, Mr Lloyd, recommends that she be sent away to Lowood School, where she excels and goes on to become a teacher. Soon thereafter, Jane takes up the position of governess for a young girl, Adèle

Varens, at Thornfield Hall, a country estate. Here Jane becomes increasingly fond of its owner, the passionate Edward Fairfax Rochester, whom she finally marries and to whom she bears a child. In the long interval between Jane's arrival at Thornfield and her marriage to Rochester, she negotiates a complex set of personalities, relationships and events that involve, *inter alia*, the mysterious Grace Poole, the so-called eccentric servant with a drinking problem who guards Rochester's deranged wife, Bertha Mason, secreted away at Thornfield; the beautiful Blanche Ingram, whom Rochester is expected to marry; the clergyman St John Rivers; and two fires, one that sets ablaze the curtains and bed in Rochester's room, and one that burns Thornfield down and results in Rochester losing a hand and his eyesight, which towards the end of the novel is partially restored.[57]

Throughout the novel the imagery of fire is collapsed with Jane's developing sense of independence and of her own sexuality.[58] Early on, in chapter Four, during Jane's troubled time at Gateshead, she contemplates her growing feelings of unease:

> A child cannot quarrel with its elders, as I had done; cannot give its furious feelings uncontrolled play, as I had given mine . . . A ridge of lighted heath, alive, glancing, devouring, would have been a meet emblem of my mind when I accused and menaced Mrs Reed; the same ridge, black and blasted after the flames are dead, would have represented as meetly my subsequent condition, when half-an-hour's silence and reflection had shown me the madness of my conduct, and the dreariness of my hated and hating position.[59]

Here fire shifts from something that animates a stretch of moorland – and therefore becomes an appropriate metaphor for Jane's independent streak – to one that turns the heath into something 'black and blasted', a phrase that captures evocatively the actual effects of fire as well as Jane's own doubts about her 'mad' methods and her forlorn circumstances.

These careful framings of fire weave in and out of Brontë's story. In chapter Twelve, Jane assists a man whose horse had slipped on a patch of ice: this is her first introduction to the enigmatic Rochester, the proprietor of Thornfield, her employer and her husband to be. It's a crucial moment in the development of the narrative. Jane says:

> Had he been a handsome, heroic-looking young gentleman, I should not have dared to stand thus questioning him against his will, and offering my services unasked. I had hardly ever seen a handsome youth; never in my life spoken to one. I had a theoretical reverence and homage for beauty, elegance, gallantry, fascination; but had I met those qualities incarnate in masculine shape, I should have known instinctively that they neither had nor could have sympathy with anything in me, and should have shunned them as one would fire, lightning, or anything else that is bright but antipathetic.[60]

In this configuration, fire is associated with the seductive and dazzling but finally the hostile. A few pages later, however, Jane warms to the cosy experience of Thornfield's hall:

> Not dark, nor yet was it lit, only by the high-hung bronze lamp; a warm glow suffused both it and the lower steps of the oak staircase. This ruddy shine issued from the great dining-room, whose two-leaved door stood open, and showed a genial fire in the grate, glancing on marble hearth and brass fire-irons, and revealing purple draperies and polished furniture, in the most pleasant radiance.[61]

The domestic fires that greet Jane at Thornfield (after the accident described above) comfort her, but no sooner has she absorbed their 'pleasant radiance', than they 'reveal' a group near the mantelpiece:

> I had scarcely caught it, and scarcely become aware of a cheerful mingling of voices, amongst which I seemed to distinguish the tones of Adèle, when the door closed.[62]

Jane hastens to Mrs Fairfax's room, where there 'was a fire too, but no candle'.[63] The concluding lines of this chapter make it clear that the man Jane had come across in the snow is none other than Mr Rochester. Thus begins their ardent relationship, one that is launched, as it were, by fire's 'illuminating' nature. And it is this fiery narrative that continues into the two blazes that provide the spectacular underpinnings to Brontë's story.

Both engineered by Rochester's hidden wife Bertha, the first appears in chapter Fifteen. Sleeping uneasily in a 'drearily dark' night that causes Jane to say that she 'wished she had kept my candle burning', she hears a 'demoniac laugh'.[64] A gurgling and a moaning follow, whereupon Jane 'hurried on my frock and a shawl' and opened her door; she sees smoke and smells burning.[65] Bringing together sounds, sights and smells, always present in the experience of fire (think back to William Sansom's essay, 'The Naked Flame', mentioned in the Preface), these passages set the scene for Jane's terrifying experience of a burning Rochester: 'Tongues of flame darted round [his] bed; the curtains were on fire. In the midst of blaze and vapour, Mr Rochester lay stretched motionless, in deep sleep.' Rushing to 'his basin and ewer', Jane 'deluged the bed and its occupant'; then, hurrying back to her room to fetch her 'own water-jug', she 'baptized the couch afresh, and by God's aid, succeeded in extinguishing the flames which were devouring it'. Roused at last (and this is the word Jane uses), Rochester exclaims, 'Is there a flood?'. 'No, sir', Jane answers, 'but there has been a fire; get up, do; you are quenched now. I will fetch you a candle.'[66]

Ignited in that most intimate of places, the bedroom, where slumber and sex habitually cohabit,[67] this fire and its descriptions lean freely on religious and secular symbolism, from the obvious allusions to

Christianity via baptism and the deluge, to the more veiled references to sexual arousal and release. In respect of the latter, Sally Minogue has observed:

> Bertha Mason may set the conflagration in Rochester's bed, and Jane may put it out, but in the aftermath passion is set alight in what is close to a bedroom seduction scene. Jane is now truly 'awake'.[68]

She is, but her final, gentle aside to Rochester, 'I will fetch you a candle', seems to return her to a more temperate condition, one in which a candle's benevolent properties, implied throughout the entire scenario described above, is always positioned as the antithesis of 'fire': it does not inflame, but rather consoles.

Rochester finally asks Jane to marry him; circumstances, however, soon reveal that the woman imprisoned in the attic is in fact Rochester's wife, who went 'mad' soon after their marriage some fifteen years earlier. Jane flees Thornfield. Destitute, the Reverend St John Rivers and his two sisters take her in. From him, Jane learns that her uncle, John Eyre, had left his fortune to her, and that Rivers and she are cousins. He will go on to propose to her in what Jane recognizes to be an expedient measure to facilitate his move to India as a missionary. Even so, she certainly does not love him, and she declines. The seeds for this denouement occur in chapter Thirty-three where, once again, the imagery of fire plays an essential role. Rivers arrives in Jane's room, his clothes covered with snow: he says, 'I shall sully the purity of your floor'. Then he approaches the fire. Jane, for a moment, is attracted to his 'handsome-featured face . . . like chiselled marble'.[69] Words pass between them; then this short, abrasive dialogue follows:

> 'But I apprised you that I was a hard man', said he, 'difficult to persuade.'

> 'And I am a hard woman, impossible to put off.'
> 'And then,' he pursued, 'I am cold: no fervour infects me.'
> 'Whereas I am hot, and fire dissolves ice. The blaze there has thawed all the snow from your cloak; by the same token, it has streamed on to my floor, and made it like a trampled street.'[70]

These quick-witted comments cut to the chase, or so it would seem. On a second reading, they reveal a nicely judged ambivalence that is entirely suited to Jane's uncertain feelings and to fire's equally conflicting symbolism that is used to describe them. On the one hand, Jane seems to be embracing Rivers's overtures by saying that her passionate ('hot') nature is able to 'dissolve' his ice-like demeanour. On the other, she is pitting her own independence, as a 'hard woman', against Rivers's ungiving ('cold') disposition. It is not until Jane draws attention to 'The blaze there' – the fireplace in her room – that her true thoughts are made clear: that 'blaze', she says, has thawed the snow from Rivers's clothes, but it has also caused it to fall on her floor and made it into a 'trampled street'. A trampling: that says it all, I think.

Jane's razor-sharp comments are bound up with her striking invocation of the contradictory images, fire and ice; they invite comparison with the American Robert Frost's poem, 'Fire and Ice', first published in 1923 in his anthology *New Hampshire*, which went on to be awarded the Pulitzer Prize:

> Some say the world will end in fire,
> Some say in ice.
> From what I've tasted of desire
> I hold with those who favor fire.
> But if it had to perish twice,
> I think I know enough of hate
> To say that for destruction ice

> Is also great
> And would suffice.[71]

As Jay Parini observes in his study of Frost, here the poet/narrator brings together an extraordinary range of emotions in remarkably few words. There are 'easy generalizations ("Some say")'; 'self-consciously elevated diction ("I hold with those")'; and the 'wryly comic, wittily rhyming "ice" with "suffice"'.[72] In summary, Parini says, 'One can scarcely imagine a more complex linking of tones in such a short space, or such compacted fury combining with such quiet, effective comedy.'[73] These comments on the formal structure of Frost's poem seem a perfectly apt description of Jane's volatile dialogue with Rivers.

And so to the final conflagration in *Jane Eyre*, which has as its preamble a startling dream, part of which Jane relates to Rivers and part of which she thinks upon privately. In the dream she hears a voice crying out to her from somewhere; it is that of Rochester; and it is one of 'pain and woe'.[74] Departing the following day for distant Thornfield Hall, where Rochester still resides, Jane eventually arrives there only to find 'a blackened ruin' and 'shattered walls'.[75] Retiring to a nearby inn, she exchanges words with its owner, who turns out to be Rochester's former butler. He proceeds to describe the blaze at Thornfield:

> The fire broke out at dead of night, and before the engines arrived from Millcote, the building was one mass of flame. It was a terrible spectacle: I witnessed it myself.[76]

This emphatic statement of personal experience serves to underline Jane's removed – but fiercely felt – realization of the fire, one that came to her, as we have said, in an ill-defined presentiment; an experience that stands in sharp contrast to her absorbed engagement in Rochester's first, terrible burning.

This second fire, we soon learn, has left Rochester blind. Considering the intense visuality of the novel up to this point, this loss of vision of a central character is significant. It is as if sight and all it has to offer is finally superseded by other senses, including, of course, love, which underpins the novel and will soon be consummated in Jane's marriage to Rochester. Perhaps to strike home this point, Brontë makes Jane's awareness of Rochester's tragic circumstances come by way of a premonition, one that, as we have seen, is located in a quiet house and a fading candle. The other 'fire', as it were, is just around the corner.

four
Playing with Fire

> The forest has caught fire
> The great forest is blazing;
> Shut your eye
> You will still see it burning.
> It burns, friends, it burns.
> Be careful with fire![1]

This is the first stanza of 'The Fire' written by the Swahili poet Ahmed Nassir Juma Bhalo, in or about 1960, at a time when the dissolution of colonization in Kenya and elsewhere in Africa matched the breaking down of cultural constructions of 'modernism' and the slow emergence of the 'postmodern'. The poem evokes in crisp language the enormity and relentlessness of a forest fire; but it is also a call to be cautious: the line, 'Be careful with fire!', will be repeated at the end of each of the remaining three stanzas. Composed against the backdrop of the Mau Mau resistance to British rule in Kenya (1953–9), it may be that Bhalo intended fire to be a metaphor for African nationalism; something that, like fire, was intrinsically volatile, contradictory and always in need of careful management. Between fact and admonition, however, he opens up an imaginary dimension through which fire, as it were, may be seen: 'Shut your eyes', he says, 'you will still see it burning'.

For many artists in the second half of the last century, fire provided a powerful catalyst for wide-ranging engagements with medium and message. The narratives of fire discussed so far continue – shifting references to history and myth; destruction and recovery; intimacy and awe. But now fire's volatile nature is reformulated in ways that locate it variously in the playful, the political and what might be called the spiritual. Many artists turn to fire itself rather than its representations, including

Yves Klein, Ana Mendieta, Gina Pane, Mike Parr, Peter Fischli and David Weiss and Jerzy Bereś. This is something palpably new, and consistent with the breaking of boundaries that characterized advanced art in the late 1950s onwards, where anything and everything could be used to make art. It is more than that, though: it is as if fire became, for a period of some two decades, the medium of choice amongst the avant-garde, whether it was in Paris, New York or Sydney. But when that 'deadly prankster'[2] – Neil Watson's memorable description of fire – is harnessed to creative activity, what are we to make of it? Is it simply another instance of human control over nature or an attempt to usurp fire's destructive role and transform it into the transformative?

Fire and Play

Play takes us into childhood games and antics where anything seems possible and where everything is capable of becoming something else. Reflecting on these ideas – the play of children, but also the play of adults – the Swiss duo of Peter Fischli (*b.* 1952) and David Weiss (*b.* 1946) immediately come to mind,[3] particularly their famous *Der Lauf Der Dinge* (*The Way Things Go*), 1987, a remarkable 30-minute film in which fire makes frequent appearances.[4] Some 25 years later, the film continues to captivate. Mimicking the small world of childhood with its toy-like objects and its fantastic effects, the work all the while brings to bear a perspective that is anything but ingenuous. In some instances, too, it hints at narratives that belie the film's (seemingly) dispassionate – and innocent – configuration.

Fischli and Weiss work across a range of mediums, from sculpture to photography to moving image. Their first collaborative project, the '*Wurstserie*' ('Sausage Series'; 1979), comprised small colour photographs showing, amongst other things, sausages on wheels (yes, they do resemble penises at play), cigarette butts, a whisk and cardboard boxes-cum-buildings, some of which are burning.[5] In other photographs, such as *The Fire of Uster* (illus. 58), these cardboard boxes-cum-buildings are shown simply smouldering. I first read the title as 'Ulster', and made the obvious connection with the province of Northern Ireland and its well-known associations with sectarian violence undertaken by the Ulster Volunteer Force in the early 1970s. This helped explain the burning structures that constitute the work. Then, realizing that the word was in fact Uster, I soon learnt that this is a small industrial town in the Swiss canton of Zurich. Further investigations took me to the home page of Büchi AG, Industrieglas, which temptingly began with the words, in bold, 'At the

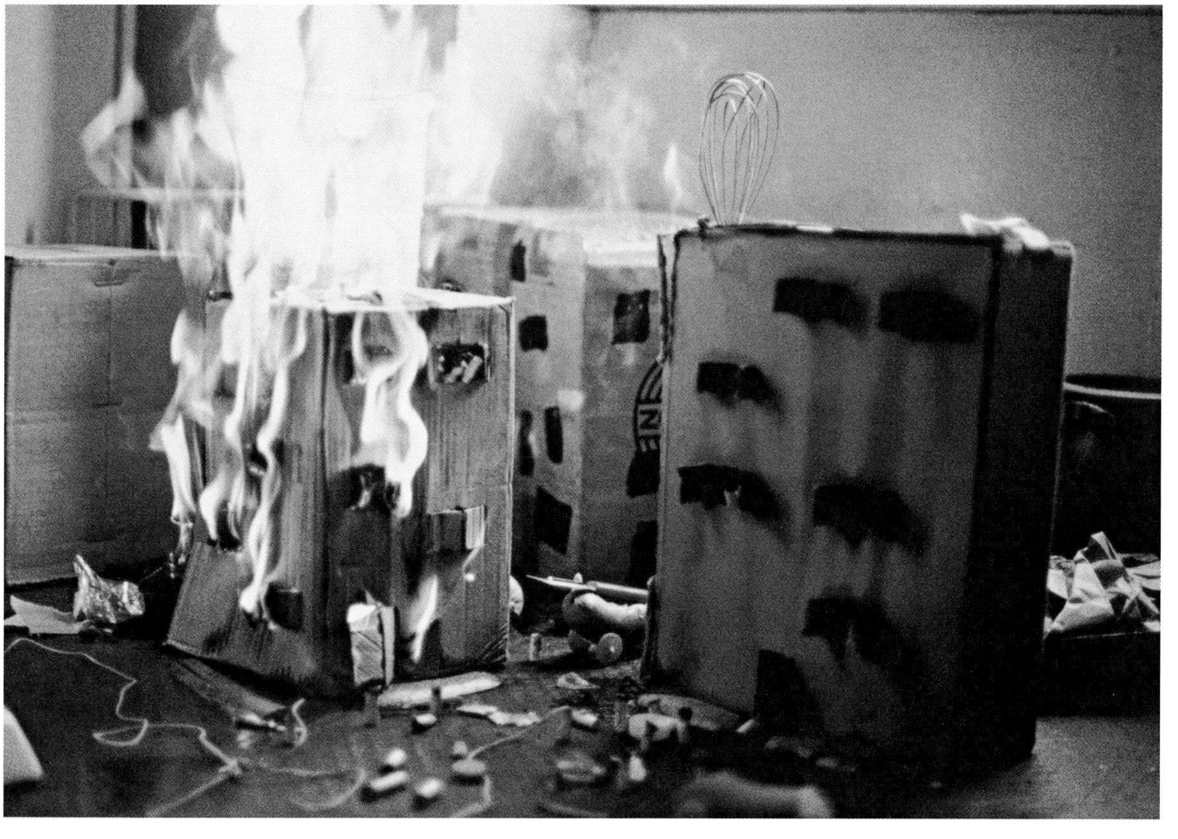

58 Peter Fischli and David Weiss, *The Fire of Uster* from the 'Wurstserie', 1979, C-print.

beginning there was fire', followed by: 'Everything started in Uster, in 1946, with the production of glass apparatus for hospitals, laboratories and industrial use.'[6] Here fire, of course, relates specifically to glass and its production. Yet there was an actual fire in Uster in 1946, on 14 November, in a *ziegelhütte* (brick factory), but this was nothing out of the ordinary, simply 'a fire like almost every year'.[7]

Meanwhile, Fischli and Weiss themselves have spoken about a fire in a factory in Uster in 1832, one perhaps started by workers.[8] All this illustrates the ways in which their art transcends its playful programme, often in spite of itself, and provides a context for *Der Lauf Der Dinge*, a work that combines parody and pastiche through technical effects that are demonstrably old hat, but ones that seem entirely appropriate to the project's (supposed) desire to demonstrate, as Virginia Rutledge says, 'the utter rationality of cause and effect'.[9] A closer viewing, however, reveals that this sequential piece (the movement of the camera appears to 'follow' the chain of events described below) is made up of a range of filmic tropes – close-ups, mid-shots, long shots and dissolves – that break the illusion that the work is an unedited sequence of cause and effect. The few stills reproduced here cannot begin to capture the work's charisma (illus. 59, 60, 61). And it is 'charismatic' as a consequence of the very human attributes that the artists bestow on their objects. There are, amongst other things, tyres without their tubes; hybrids that resemble marvellous toys (like those of '*Wurstserie*'); candles; an assortment of ramps; a Bauhaus-type kettle; balloons, bottles and tins. Big and small eruptions of fire, some recalling the fun of fireworks, pepper the film, as do the sounds of the gurgling and dripping of liquid: two elements constantly vying with one another for the viewer's attention.

The film opens with an ill-defined image soon revealed to be a large black plastic bag revolving slowly above a tyre; touching it – finally – the bag sends the tyre moving; it hits one tyre then another; a bottle is struck in the process and falls off a table: was this intended? Things continue: tubeless tyres and bottles meet and separate; a soft, grey-white substance

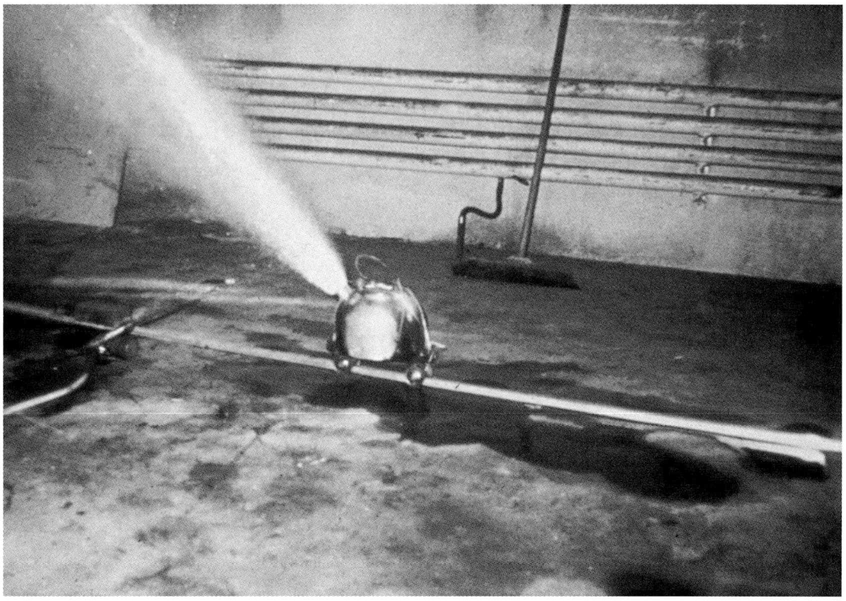

59 Peter Fischli and David Weiss, still from the film *Der Lauf Der Dinge* (The Way Things Go), 1987.

oozes out and around two very small candles, and suddenly bursts into flame. 'Cracks' and 'pops' set the ambient background. And on it goes. Structures fall; the slowly moving, glacial-like liquid-cum-gas cascades over a table; and a carpet unfurls. On what looks like a white canvas, a small ignition occurs, sending flames running along in slithery lines across its surface: in its unfolding configuration, it brings to mind the rough outline of a body; an image that recalls the Cuban artist Ana Mendieta's '*Silueta*' ('Silhouette') series from the mid-1970s. Another candle is lit, a toy vehicle is propelled down a ramp, and soon a close-up of a tyre, slowly burning, fills the screen: all now is flames and burning rubber.

Menacingly, this brings to mind burning 'necklaces', the euphemism given to the practice widespread in Apartheid South Africa in the 1980s of placing a tyre soaked in petrol around the victim's neck and then setting it alight: the first extensively reported 'necklacing' took place in the

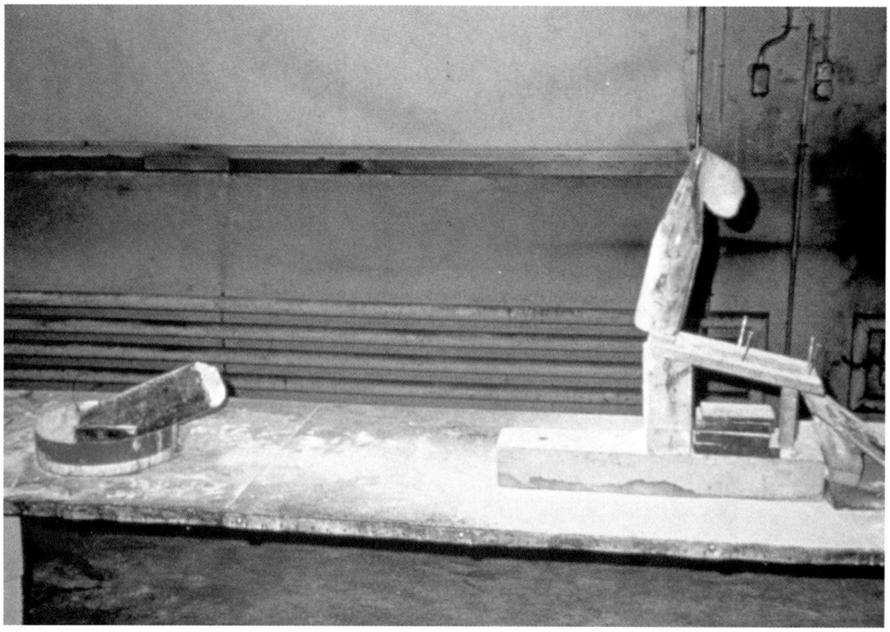

60 Fischli and Weiss, still from the film *Der Lauf Der Dinge*.

Eastern Cape in March 1985 and involved Mr Tamsanqa Kinkini, a member of the town council of KwaNobuhle, who had been accused of involvement in violence and corruption; the next burning to receive widespread coverage was that of a young woman, Maki Skosana, in July 1985, in Duduza township neighbouring Johannesburg. Skosana, who was accused (incorrectly, as it would later turn out) of being an informer, was drenched in petrol and burnt to death. The horrifying practices of burning, and in more recent times of 'necklacing', have been examined by Joanna Ball who locates these forms of summary punishment in the broader context of fire, ritual and purification in African societies.[10] The initial burning of witches in rural communities took place at times of great social unrest and such conditions, Ball suggests, were to be found in urban townships in the mid-1980s when 'necklacing' gained notoriety; she also draws attention to Bantu notions 'that fire can destroy evil, that it can purify'.[11] And she comments further, and provocatively, on the ways in which the visual fascination of fire itself might supersede the significance of the tyre, although considered together, the symbol of the tyre/necklace is a powerful one. Such observations point once more to fire's paradoxical make-up, in this instance, to a mixing of emotions ranging from fascination to fear together with more symbolically loaded ideas of redemption and purification.

I am not implying that Fischli and Weiss reference the practice of necklacing directly, although one feels sure that they would have been aware of it. Between the beginning of 1984 and mid-1987 (during which time *Der Lauf Der Dinge* was made), some 672 people were burnt in South Africa, half of whom were 'necklaced'. What I am suggesting,

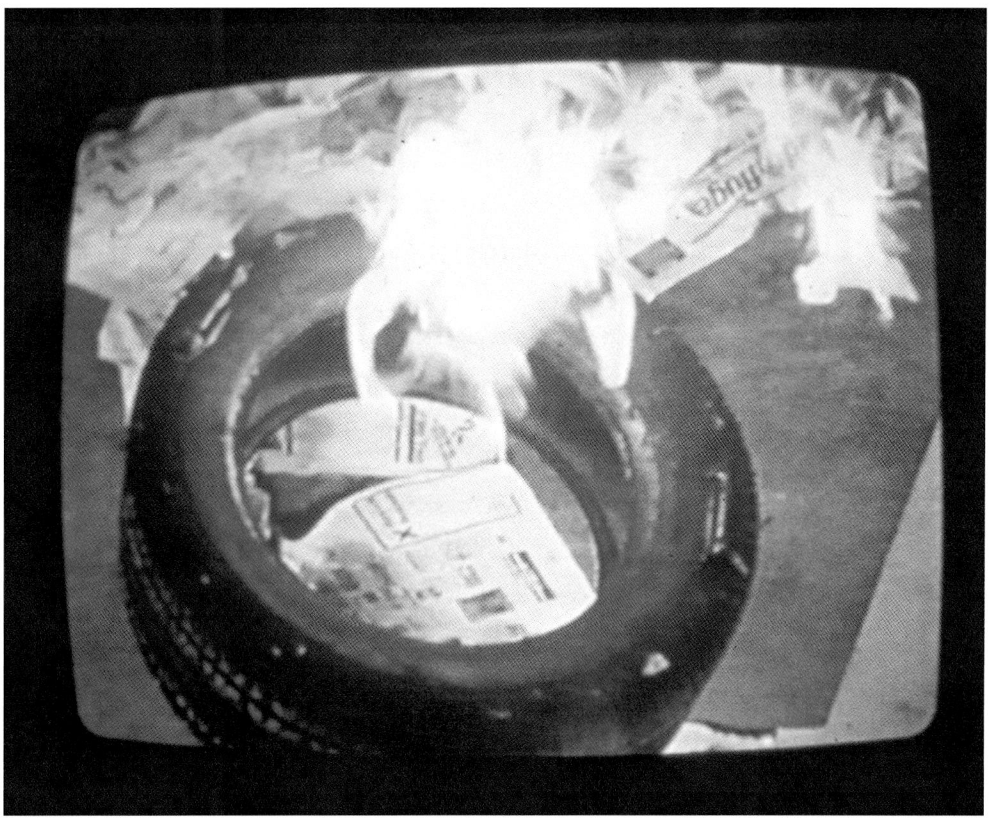

61 Fischli and Weiss, still from the film *Der Lauf Der Dinge*.

however, is that Fischli and Weiss's oeuvre can easily segue from play into matters more weighty – which is not to say, of course, that play *per se* precludes the 'serious'. The last few minutes of *Der Lauf Der Dinge* reintroduce some of its chief motifs and effects: there are the tyres and the discharging liquid/gas and, of course, the flames, moving along surfaces and dropping, liquid-like: the fire paintings of Yves Klein (discussed later) spring to mind. *Der Lauf Der Dinge* concludes where it had begun, but not quite: the large black plastic bag is there, but now wrapped in rope; a Christo/Jeanne-Claude-like object that finally falls, almost unwillingly, into a tractor tyre, after a chain of events that collapse structures. I suspect this latter result was not intended, but regardless, it releases the gaseous liquid that now fills the screen and leaves the viewer with a final, tender image of hair-like lines emerging, very briefly, out of the vapour.

Der Lauf Der Dinge is a mesmerizing mix of odds and ends and effects that, like a child's understanding of the world, has its own strange inner logic. Yet, of course, it is a carefully planned and executed work.

By contrast, Jean Tinguely's *Homage to New York* (1960), in which fire also plays an important, albeit unintended, role, was from beginning to end a work-in-progress, where the accidental and the premeditated constantly rubbed shoulders.[12] Tinguely was one of the original members of the Nouveaux Réalistes, a group of French artists that the writer and art critic Pierre Restany formally founded on 27 October 1960 (and which included, amongst others, Yves Klein, César, Arman and Jacques Villeglé). Tinguely was always attracted to ideas of movement and, therefore to the machine itself. Not the latter's association with efficacy, but rather the converse, its potential to demonstrate a self-sufficient absurdity, doing everything and yet doing nothing.[13] Tinguely plays with this notion in much the same way that Fischli and Weiss play with their broad range of objects. Theirs, as I have said, is an oeuvre that always hints at other narratives. Not so with Tinguely, where what we see is largely what we get. *Homage to New York*, however, is one of the few exceptions. Like *Der Lauf Der Dinge*, it is an 'event' that we know through film and photographs. Tinguely's had an official opening in the sculpture courtyard of New York's prestigious Museum of Modern Art on 17 March 1960, while Fischli and Weiss's 'incident' was filmed in the private space of a warehouse.

Contemporary descriptions of *Homage to New York*, published in the *New York Times* in 1960 and, two years later, by Calvin Tomkins in *The New Yorker*, describe it in detail. Tomkins's is widely recognized as the most authoritative and needs to be quoted at length, specifically in regards to the artwork's haphazard genesis and its (unintended) deployment of fire and its (unexpected) results:

> George Staempfli, a New York gallery owner . . . had met Tinguely in Paris and promised him a New York show . . . 'It has to be in the Museum of Modern Art' [Dore Ashton, an art critic, remembers Tinguely saying] . . . The machine that Tinguely envisaged . . . [would] destroy itself in one act of glorious mechanical freedom . . . He bought dozens of used motors and a powerful electric fan . . . In an Army-surplus store he found a big orange meteorological balloon and some signals. His friends collected enough money for him to buy a quantity of steel tubing . . . and the museum provided him with his indispensable tool, an acetylene torch.[14]

Tomkins goes on to describe how Billy Klüver, a young Swedish research scientist working at the Bell Laboratories in New Jersey, became Tinguely's assistant. Klüver purchased over 30 old bicycle wheels, which would prove to be crucial in the work's final design. 'When Jean saw them', Tomkins writes, 'he was like a child . . . His ideas took shape from them. That's how he works. He creates with what comes to hand, and the material shapes the creation.' As the piece developed, various other odds and ends entered Tinguely's armoury, including an oil-can, a cable drum, a piano, a child's bassinet and an old addressograph machine. 'Because the show was scheduled for early evening, Tinguely decided at the last moment to paint the entire construction white, in order to make it more clearly visible.' Tomkins captures the hectic last-minute preparations: the New York artist Robert Rauschenberg brought a 'money thrower', a device that was calculated to throw up silver dollars once ignited by flash powder; Klüver 'brought several containers of titanium tetrachloride, for making smoke'; assistants were running about here, there and everywhere.

And so to *Homage to New York* itself. It was by all accounts a comedy of errors, the antithesis of *Der Lauf Der Dinge*:

> Klüver put in the plug. The machine gave a shudder, rattled and immediately broke down. The belt driving the piano mechanism had slipped off its wheel as soon as the motor started . . . The piano started to play again but other belts had jumped off and only three notes were working . . . What happened next was pure, vintage Tinguely. Thick, yellow, strong-smelling smoke billowed from the baby's bassinet, where Klüver had put the titanium tetrachloride . . . spectators now found themselves enveloped in a choking cloud . . . Smoke and flames began to emerge from inside the piano . . . The radio turned itself on, but nobody could hear it . . . The meteorological balloon refused to burst . . . The piano was really blazing now: 'There is something very odd about seeing a piano burn', George Staempfli has since said. 'All your ideas about music are somehow involved.' For the museum authorities, a good deal more than ideas about music was involved. They had not anticipated a fire, and were understandably sensitive on that subject in view of the museum's second-floor fire the year before.

Tomkins concludes his account with a vivid description of the developing fire in the piano:

> The concealed fire extinguisher [in the piano] was supposed to go off at the eighteenth minute, but the flames had spread through the whole piano and burned out a vital connection . . . Some twenty minutes after the start of the destruction, the resistors in the sawed-through supports began to melt the joints . . . Tinguely suddenly began to worry for fear the fire extinguisher in the piano would explode from the heat . . . Klüver found a fireman in attendance . . . with apparent reluctance [the firefighter] signalled to two museum guards, who ran out with small extinguishers and applied them to the piano fire. The audience booed the men angrily, assuming that they were spoiling the show . . . More boos, interspersed with a few cheers, as the fireman and the two guards attacked the sagging machine with axes, knocking down the piano, the big meta-matic, and the mast that held the balloon. At the end, though, there was a rousing cheer for the dying monster.

Tomkins wrote his article some two years after the event. While there is surely embellishment here and there, it nonetheless reveals *Homage to New York*'s beginnings and outcomes to be a muddled affair in which play is ascendant. Tinguely 'plays' like a child; together with Klüver, he rearranges objects into new configurations and derives pleasure from the unexpected results. 'Many people told me afterward', he would say,

> that for them it was a terrible thing. One woman told me later that she had wept, she found it so sad. It was sad, in many respects. It was also funny. I laughed a great deal while it was happening. I was the best spectator.[15]

In all of this, fire took on its role as 'dreadful prankster': unexpected, bursting out here and there, potentially alarming but, finally, and in comic circumstances, brought under control by a firefighter and two museum guards.

I have not illustrated Tinguely's self-destructing machine. I do so now. It's not the actual work, though, but a drawing of one of its (unfolding) components, the burning piano, by Michael Landy, from a recent series he exhibited at Alexander and Bonin, New York, called *Michael Landy* H2NY (*Homage to New York*; illus. 62).[16] Works from this show were recently seen at Sherman Galleries, Sydney in 2007, but now with a new title, *Man in Oxford Street is Auto-Destructive*.[17] One of the celebrated YBAs (Young British Artists) from the late 1980s, Landy is best known for *BreakDown*, 2001, a work in which he destroyed his possessions (the BBC documented the event in a film called *The Man Who Destroyed Everything*, dir. Nadia Haggar, 2001). For the project, Landy

> took over the vacant C&A shop in London's Oxford Street and set up an airport-type carousel that delivered all his worldly goods into the hands of a team of helpers who pulled each object apart, limb from limb.[18]

Landy's typically cryptic and ironic voice resonates in the title for his Sydney exhibition, 'Man in Oxford Street is Auto-Destructive'. Sydney, too, has a street of the same name as London's, located very near to Sherman Galleries; it is a street notable for its cafes, boutique shops and markets, the antithesis, it would seem, of Landy's homage to Tinguely. But perhaps not. Consumption, as the director Antonioni showed most powerfully in the concluding sequence of his film *Zabriskie Point* (1970), may end up in a cataclysmic explosion.

I saw Landy's drawings of *Homage to New York* at the Sherman Galleries. In addition to these carefully composed and visceral works, there was also the (continuous) screening of Tinguely's artwork, a black and white film shown on a small monitor, and placed at the entrance to the gallery. Nearby, a large photograph of Landy greeted the visitor. This photograph is not what we might initially think. For the artist has shrewdly superimposed his head on the body of Tinguely, the latter dressed in

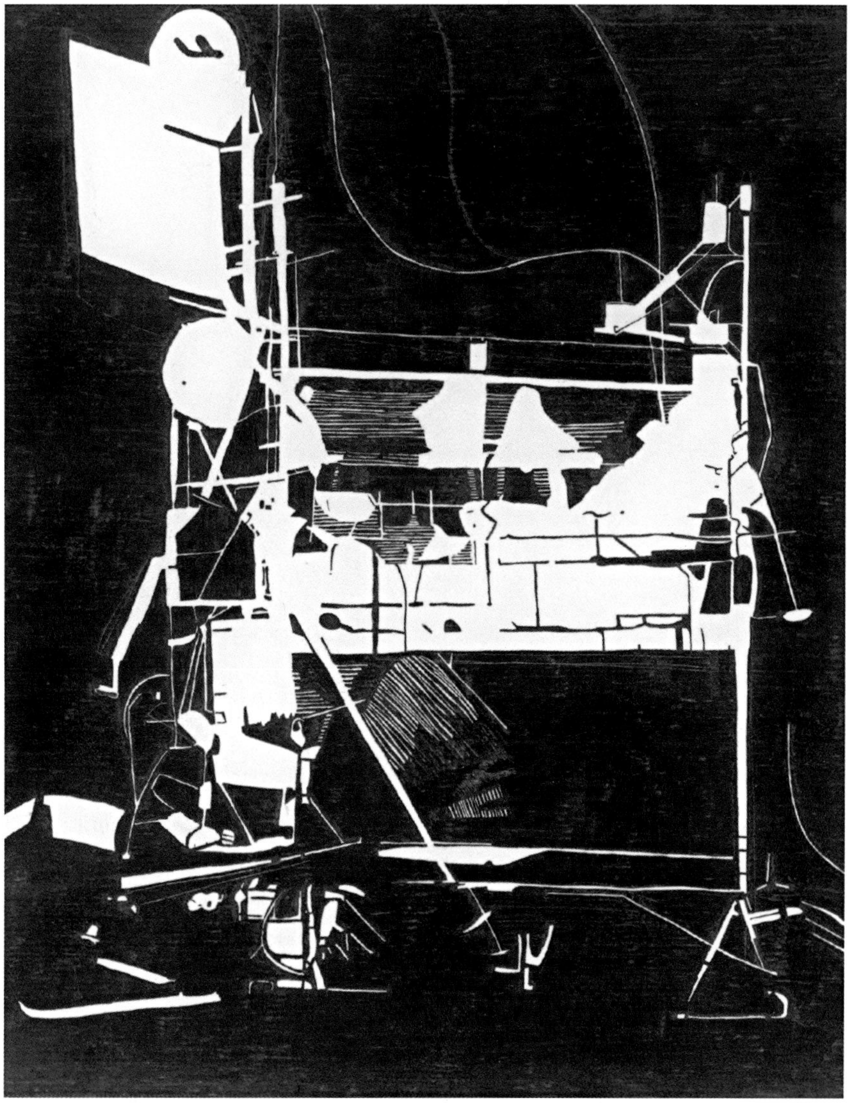

62 Michael Landy, *H2NY, The Piano Played the Last Will and Testament of a Dying Piano*, 2007, oil stick on paper.

the same way as he was in the film documenting *Homage to New York* (illus. 63). Landy looks out at us with a grin; he seems happy as Larry!

Fischli and Weiss, and Tinguely, treat fire as the ephemeral thing that it is, and all the while bring to it a note of play and childhood exuberance. Martin Honert, born in Bottrop, Germany in 1953, addresses similar themes in his three-dimensional work *Fire*, but here the transience of fire is transformed into a frozen moment (illus. 64). 'Childhood is a theme for me', he has said, 'but not because I think my childhood was especially eventful, or bad, or good . . . I think it's important to discover

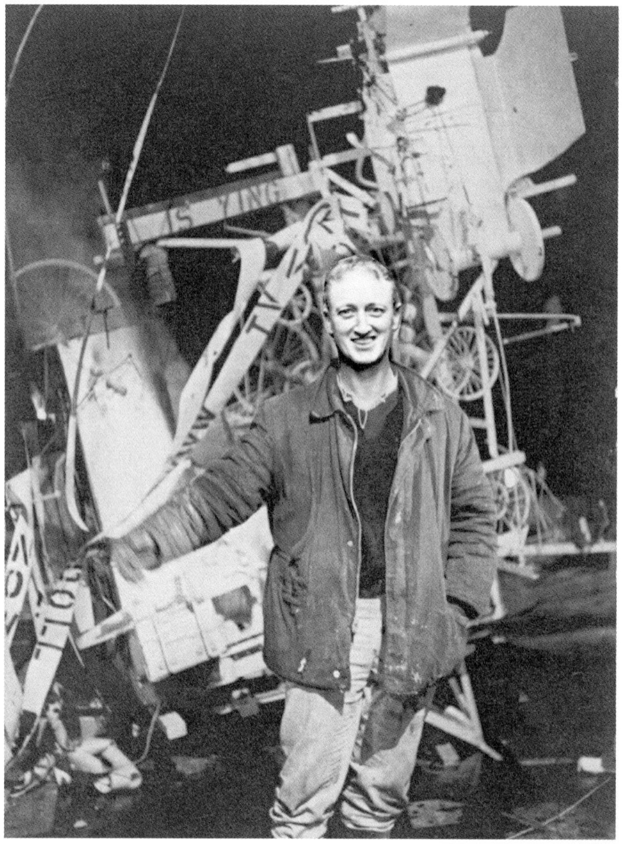

63 Michael Landy, *Man in Oxford Street is Auto-Destructive*, Sherman Galleries, Sydney, 19 July – 12 August 2007.

what's way past but still in the memory as an image.'[19] This is especially evident in Honert's paradoxically titled *Foto* (1993), a sculpture in oil and acrylic on wood, and epoxy resin, of a little boy sitting at a table: dwarfed by it, he looks out at the viewer, his expression a combination of knowingness and anguish.[20] The work is based on a slide from Honert's parents' collection, one showing the family sitting around a table while on a holiday in Spain in 1958. *Foto* confuses genres of representation: 'As you approach from the front', writes Mark Pimlott,

> the scene seems two-dimensional, as if cut out from a photograph. It is at this moment that associations flood into the work – through

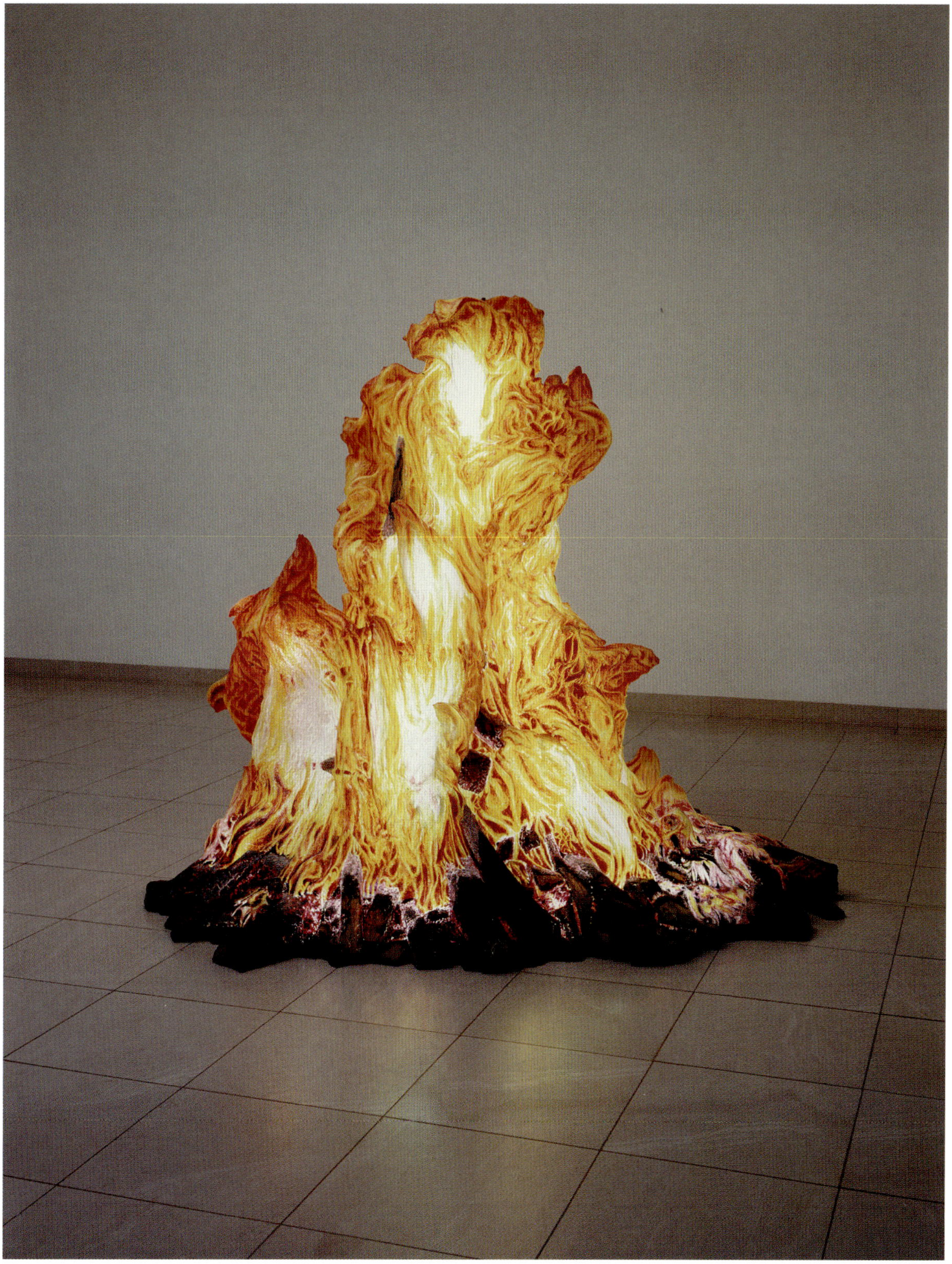

64 Martin Honert, *Feuer* (Fire), 1992, oil on cast polyester resin with fibreglass and neon.

> a convention of representation which is not that of a three-dimensional object. It is that of a family snapshot, a picture type which is intensely suggestive of the private world and its experiences, yet familiar to everyone.[21]

Fire similarly suggests this snapshot quality. Made from transparent cast polyester resins painted with glazing oil paints, a light illuminates the sculpture from within. It is a straightforward work, really, something that could be found on a stage-set, for example, or in some fast-food outlet where it might advertise the 'flaming' nature of the establishment's special fried chicken. Representing, as I have said, the antithesis of fire's transitory nature by its 'freezing' of flames, *Fire* does not show the comforting fire of the hearth or the affective nature of a bonfire. Nor is it the fire of destruction, although the piece does hint at the sinister through a suggestion of flames turning into flesh, but in the end it seems to me all about play: fire as the big and the bright and the fabulous, as a child might understand it.

Interviewing Honert, the academic Boris Groys asked:

> Looking at your works, I have the feeling that these three-dimensional pictures so objectively quoting reality . . . suggest also a certain feeling of danger, of being lost. Isn't that an allegory of the contemporary artist, who removes himself from the life context and puts himself in the institutional context of art?

Honert debunks this highfalutin interpretation, and says:

> That's not the reason I do these works. I very much like the images in dictionaries, where, for example, under the word 'fire' there's a little picture of a fire. That's actually an image in the clearest sense of the word: so, that's what a fire is! Those pictures always seem alone, isolated, removed from their context. And that's what

> interests me about them – this hermetic quality of being closed in on themselves.[22]

Text and its relationship to images and objects have long fascinated the distinguished Nebraskan artist, Ed Ruscha (*b.* 1937), now working variously in Venice, California and the Mojave Desert, and René Magritte, the Belgian Surrealist (1898–1967). Various commentators have recognized this relationship. Writing in 1971, the art critic David Bourdon went so far as to describe Ruscha as 'a sort of cowboy Magritte gone Hollywood', an association Ruscha would dismiss as too crude. 'I was more influenced by Duchamp and Kurt Schwitters and Max Ernst', he says,

> but I recognized Magritte as a powerful figure. I first saw his work in a book on surrealism in the Oklahoma City library. I just sort of stumbled across it. I guess I didn't respond to his work as much as I did to the other artists. Only later on did I see something simpatico there, something that I might have shared with him.[23]

This 'simpatico' is to do with their calling into question the nature of visual representation. Ruscha points to this in an early work, *Damage*, where the word itself, in upper case letters, spreads out the across the canvas: it touches its edges, and has a cartoon-like flame emerging from the letter A (illus. 65). Depicting the word on fire, 'the flames', Kerry Brougher writes, 'seem about to burn a hole in the issue of flatness, opening up painting to new possibilities'.[24] Here Brougher touches on Ruscha's hesitant relationship to advanced painting and its paradigms in the late 1950s, the latter most forcibly articulated by the American art critic Clement Greenberg, who argued that painting's history was all about a move to 'flatness'; a trajectory in which painting became, according to Greenberg, a self-referential medium that critiqued its own unique processes.

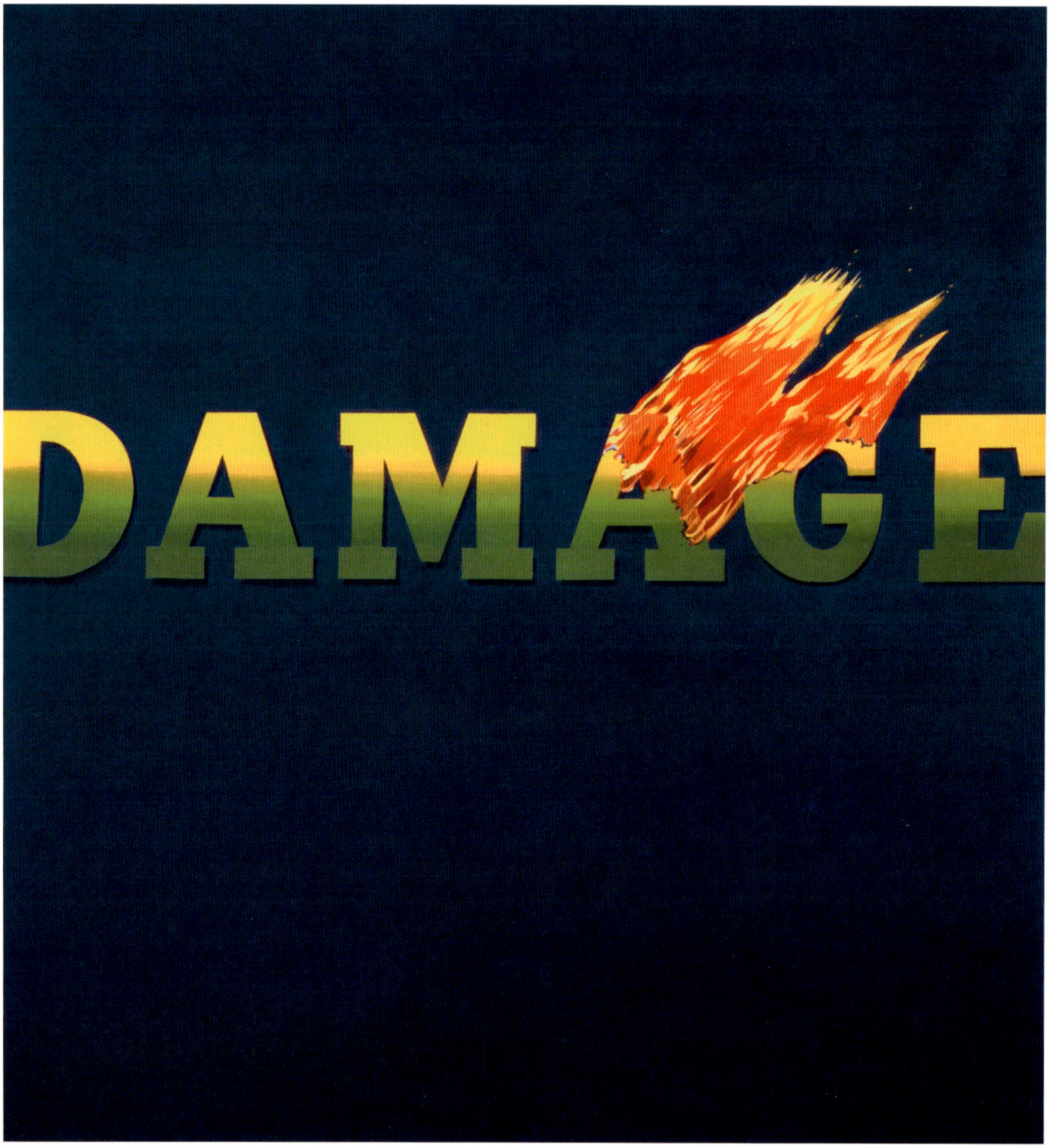

65 Edward Ruscha, *Damage*, 1964, oil on canvas.

From the early 1960s to 1975, Ruscha continued to paint, but his many other projects suggested very different avenues. This he made explicit in 1963, when he published *Twentysix Gasoline Stations*, the first of sixteen books that would variously involve photography and drawing and in which images and their relationship to text took on ambiguous relationships.[25] One of these books would be *Various Small Fires* (illus. 66),

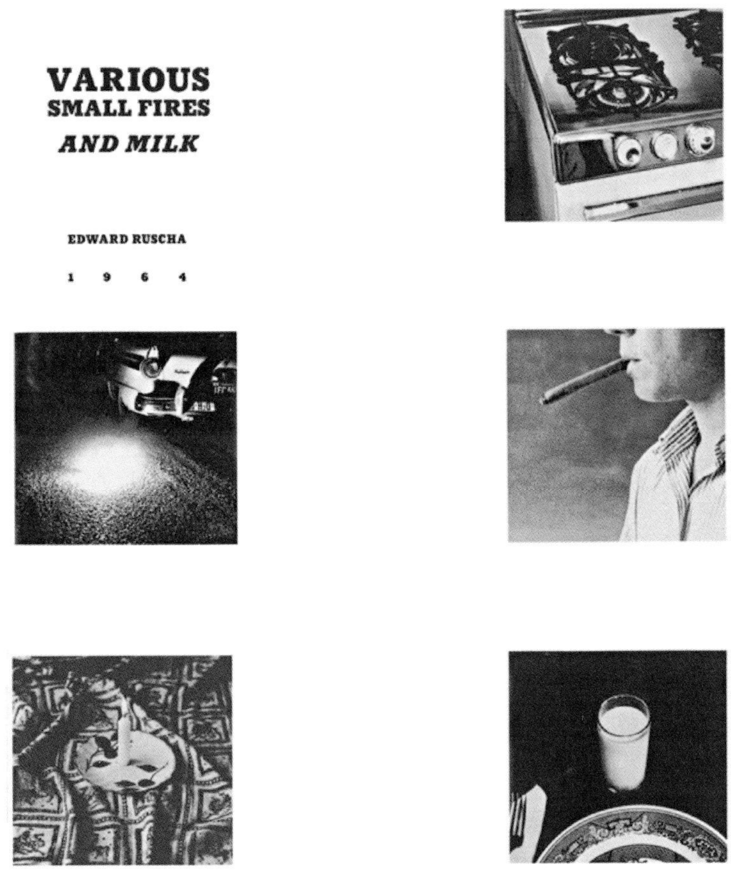

66 Edward Ruscha, composite of 5 images, along with the title page, from Ruscha's *Various Small Fires and Milk* (1964).

a teasing title, where he presented sixteen photographs, each facing a blank page. Fifteen of these are of 'small fires', literally. Isolated from any explanatory context, these tiny images seem unequivocal in their significations: they are what they appear to be, simply 'various small fires'. Yet, in their bluntness and their self-consciously 'photographed' quality – because of their frequently cropping – they convey a sense of play that makes us think twice about fire's shifting functions and meanings: a hand moves into one photograph, touching tenderly a Zippo lighter; a man smokes a long cigar, the light and shadow falling on the cigar echoing the stripes of his collar; a tongue-like flame, frozen, rises from a flip-back card of matches; and a gas cooker, is simply that – a gas appliance. The sixteenth and final photograph is of a glass of milk: three-quarters full, it is part of a table setting. This is a wry, visual *coup de théâtre*. The cover of Ruscha's book, as I have said, is *Various Small Fires*. The title page, however, reads *Various/Small Fires/and Milk*. The use of these line breaks is particularly interesting; it sets up a tension between word and image. And then there's the 'and Milk'; a liquid, of course, but not water, which so often is associated with fire and its nemesis). Why, further, should a glass of milk appear at a smart table arrangement? Considered in the light of the series, this image throws up questions of sight, site and sense. It also takes us into the comforting world of the infant, where milk is an essential ingredient to the development of strong bones and teeth. Milk nurtures, literally, and may be compared with fire's more gentle and consoling affects, but unlike the latter, it has, of course, no threatening qualities.

Ruscha's *Damage* and *Various Small Fires* characterize his cool, understated approach, one that is well illustrated in his *Los Angeles County Museum on Fire*, a large painting commenced in 1965 but not completed until three years later (illus. 67). Denuded of human presence, the building is resplendent in its modernist forms but for the discharge of fire – and 'discharge' it is – that appears at the back of the museum and seems to propel the structure forwards, not unlike the blast of fire and fumes that send vehicles into space. The building, however, stands fast, seemingly immune (or unknowing) to the goings-on at its rear. That this painting was produced at a time when the Los Angeles art community was expressing misgivings about the museum and its architecture might be significant. Yet, as Mary Richards has pointed out in her monograph on Ruscha, the fire invites a broader range of readings, ranging from the anti-institutional (the museum system), to the Watts race riots of August 1965 and to natural disasters including fire that habitually range through Los Angeles (touched on at the beginning of this book).[26] Ruscha, intriguingly, has spoken about his fascination with the nineteenth-century Pre-Raphaelite artist John Everett Millais's *Ophelia* (1851–2), which he had often seen in London at the Tate Gallery, as a major inspiration:

> I based it on that. *Ophelia* is an oblique, aerial view, and it has a very quiet, serene surrounding to it. So that painting of mine is my *Ophelia*.[27]

Considered in these terms, Ruscha's engagement with fire in *Los Angeles County Museum on Fire* situates it in a world of hard-edged architectural modernity, on the one hand, and an 'aerial' watery view (the 'serene surrounding' Ruscha ascribed to the Millais), on the other. Typically, fire here plays out its Janus-like role.

There's a lot of irony in Ruscha's work, and this, of course, assumes an initiated audience. It seems to me, however, that *Los Angeles County Museum on Fire* and other paintings such as *Sin-Without* (1991), or *The Mountain* (1998) – the latter of which depicts a mountain peak covered with snow, and the word 'The', with the capitalized 'T' in white, emblazoned at the base – may be approached simply on the basis of paintings that amuse themselves with questions of representation, making us think twice, without necessarily being aware of other agendas, about what we see and what we read and what might afford us (visual) pleasure.

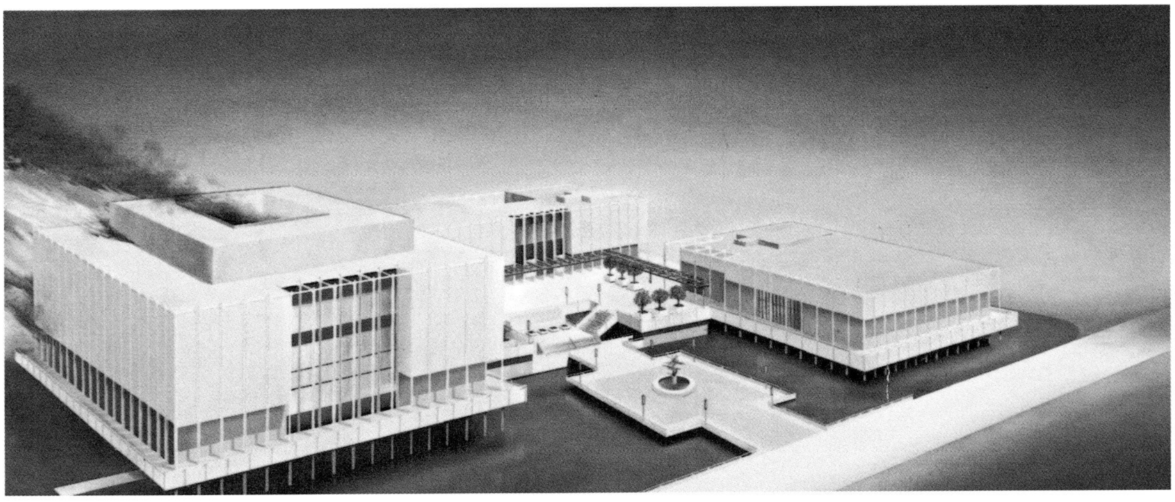

67 Edward Ruscha, *Los Angeles County Museum on Fire*, 1965–8, oil on canvas.

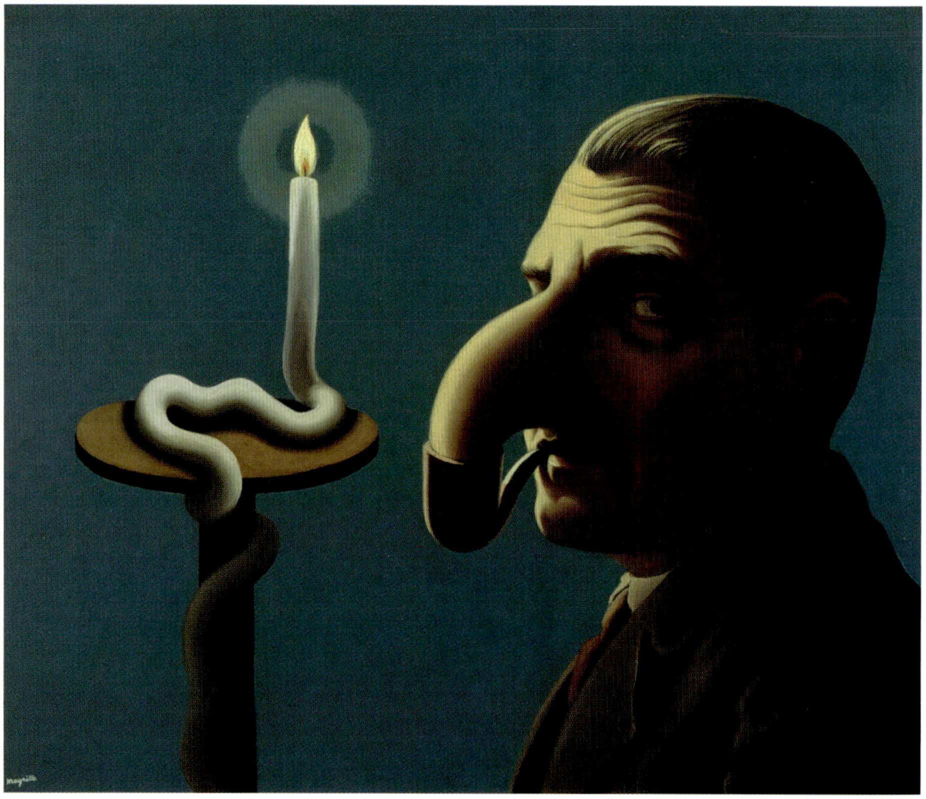

68 René Magritte, *La Lampe Philosophique*, 1936, oil on canvas.

And so on to Magritte, whose lifelong goal was to create unexpected relationships between words and objects.[28] More outlandish and calculated in his imagery than Ruscha, Magritte depicts all kinds of subjects in his art, from bowler hats, baguettes and domestic interiors, to mermaids and combs; he creates in the process a wondrous world of juxtaposition and the irrational in which fire is everywhere (he produced at least twenty works on this theme between about 1927 and 1950). A locomotive steams out of a fireplace in *Time Transfixed* (1938); a man's nose grows into a Pinocchio-like appendage and inserts itself into a pipe, while a candle nearby morphs out of a slug-like thing (illus. 68); and, most famously, *The Treachery of Images* (1928/9), a straightforward painting of a pipe,

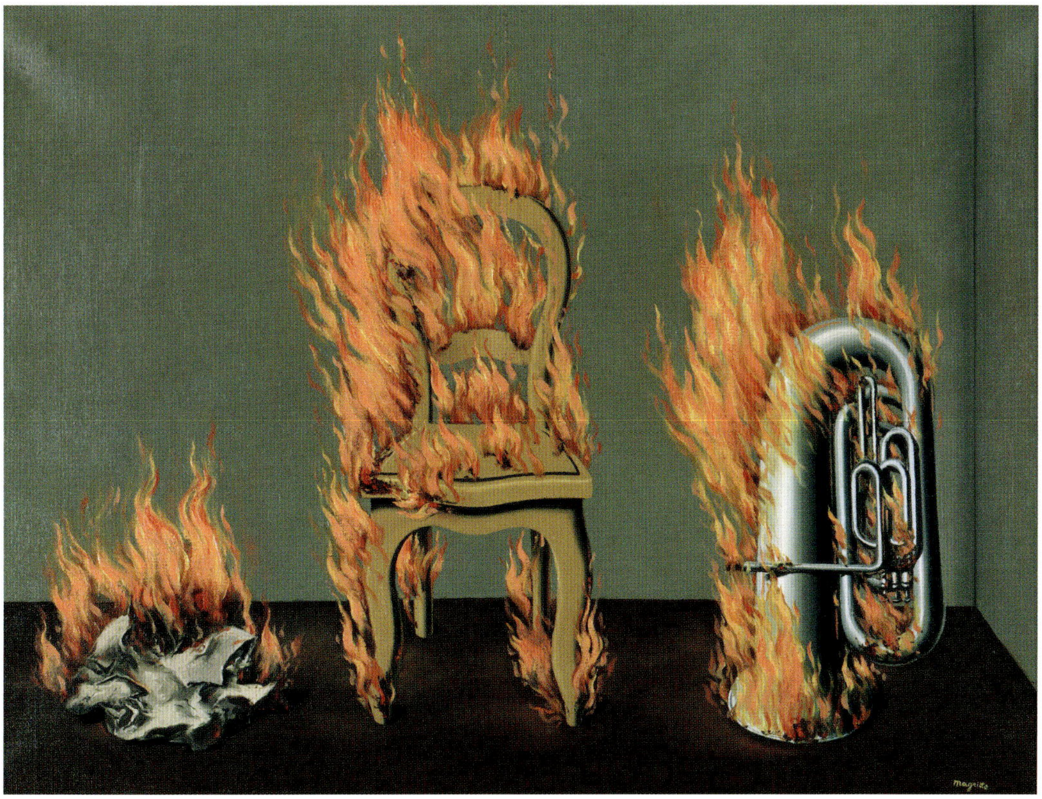

69 René Magritte, *Ladder of Fire*, 1934, oil on canvas.

Magritte has written underneath it, '*Ceci n'est pas une pipe*', a phrase that the French philosopher, Michel Foucault, would later use as the title for his book critiquing received ideas of representation.

Magritte's art typically calls into question the normative nature of language *vis-à-vis* objects and their visualization, a quintessentially postmodernist trait (Magritte was a postmodernist *avant la lettre*), Recognizing this, he takes fire into amusing and surprising situations. In addition to those paintings already cited, there is *Léchelle du feu* (*Ladder of Fire*) where he shows three objects, each made of different materials and engulfed in flame: a piece of paper, a period chair, but also a tuba. They appear in a simple room, which miraculously is immune to the fires (illus. 69). Writing to André Breton, Magritte observed in his characteristically pokerfaced manner: 'A slightly different solution would be to present a single burning object provided it was made of iron, a key, a sewing-machine or a trumpet, for instance.'[29] The trumpet soon appears as the solo image in a painting made at around the same time (1934/5), *La découverte du feu* (*The Discovery of Fire*; deleted on the verso is *L'Accord Secret* or *The Secret Agreement*). The key also turns up in a small work on paper but in the company of an egg and some crumpled paper: all ablaze, they are depicted on a tabletop seen slightly from above.[30] These works, it would seem, constitute in part a wily take on western European traditions of the still-life (*nature morte*), where objects are shown frozen in time. By contrast, Magritte gives us objects that are in the process, as it were, of self-immolation.

Finally, to Magritte's paintings that deal with fire and the motif of the cave. Of course, this association points to Plato's famous allegory, in which prisoners in his subterranean world can see only the shadows thrown by puppeteers and their puppets on the wall in front of them. The source of these shadows is the great fire behind the puppeteers. The prisoners perceive the shadows as the real; in other words, they mistake appearance for reality. Magritte would have relished this. In *Le paysage en feu* (*The Burning Landscape*, 1928), feather-like flames frame a landscape with puffy-white clouds; here the viewer is caught up in an enclosed space, looking out.[31] In *L'œil de la montagne* (*The Eye of the Mountain*, 1928), a circular, eye-like form, surrounded by flame-like tongues, hovers above a landscape of boulders; the 'eye' telescopes our attention to the indeterminate view beyond.[32] A work whose whereabouts is unknown, *La condition humaine* (*The Human Condition*), has Magritte spelling out his intentions in a letter to Paul Eluard. (The painting in question references Magritte's canonical picture of the same name, from 1933, in which he shows an easel and canvas placed in front of window; the ambiguity arises from the scene painted on the canvas: is it a representation, or is it the real view seen from the window, or might they just be one and the same?). Magritte writes:

> I am also re-doing a variant of *The Human Condition* but seen this time in a cave such as smugglers might use; in a corner of the cave there will be a fire and the landscape will be mountainous; the picture standing on the easel will show the continuation of the landscape and an old castle on the edge of a precipice.[33]

Related to these images are Magritte's series of painted bottles, at least four of which call attention to fire (the one represented here, however, does not show this element; illus. 70). Dating these works is problematic; some are in private collections while the location of others is not known. In each of the so-called 'Fire-bottles', however, the small fire seems to be in a cave 'which opens onto the starry sky', with a slither of moon appearing towards the top of the bottle.[34] The shape of each bottle differs, from the rather squat to the more elegant; one of the four has a cork just protruding. These items put painting to the service of objects, as it were, where the curvature of the container – cave-like, to be sure – dictates the design. That they are wine bottles, however, lends a mischievous slant to Plato's high-minded critique of reality; intoxication may lead to all sorts of realities.

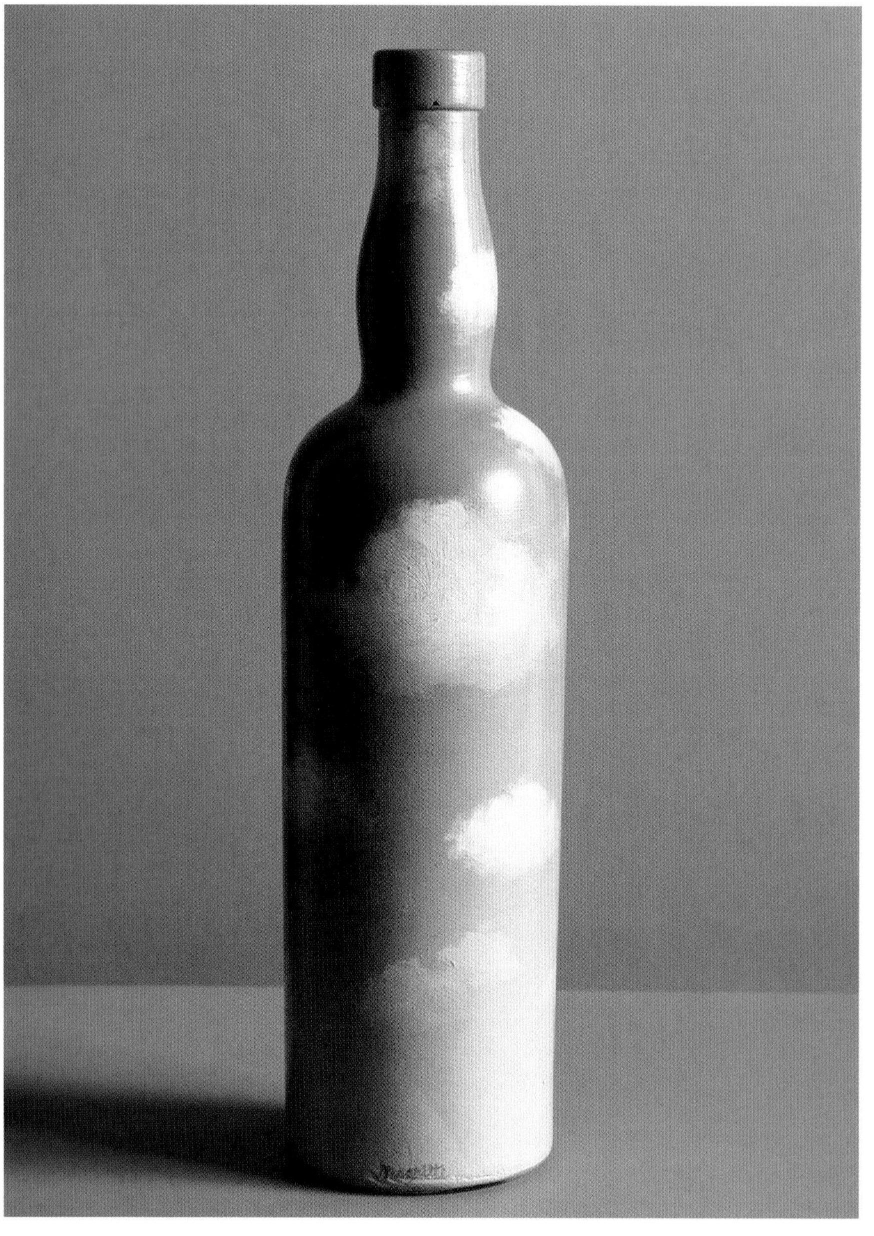

70 René Magritte, *La Courbure de l'Univers*, 1950, oil on glass bottle.

Bodies of Fire

Yves Klein (1928–1962) together with Tinguely, discussed earlier, were two of the original members of the Nouveaux Réalistes, the group of artists brought together and named by Restany in 1960.[35] Mercurial like fire itself, Klein was both showman and shaman (a catchphrase, incidentally, also applied to his contemporary, the German artist Joseph Beuys, and similarly to good effect): part philosopher and visionary; convert to the Japanese Kata (the abstract movements in Judo) and holder of a black belt of the 4th Dan; fascinated (for a while) by Rosicrucianism and also alchemy; and deeply influenced by the writings of that singular philosopher of fire, Gaston Bachelard, for whom fire's dichotomies were essential to its understanding, and to whom Klein made special mention in his (celebrated) lecture at the Sorbonne, Paris, on 3 June 1959, called 'The Evolution of Art Towards the Immaterial'.[36]

Throughout Klein's brief career – he died of a heart attack at the age of 34 – fire would play a central role in his meditations and practice, as it would in the work of the equally short-lived Ana Mendieta (1948–1985), whom I discuss below, but for very different reasons. In a lecture titled '*Le Kitsch et le Corny*' Klein delivered at the Chelsea Hotel, New York, in 1961, he famously declared:

> One should be like untamed fire in nature, gentle and cruel; one should be able to contradict oneself. Then, and only then, can one truly be a personified and universal principle.[37]

As he would do on other occasions, Klein here freely borrowed from Bachelard's *The Psychonalysis of Fire* (1938), notably from the introduction to chapter one, where Bachelard writes, *inter alia*, that of 'all phenomena', fire

> is really the only one to which there can be so definitely attributed the opposing values of good and evil . . . It is a tutelary and a terrible divinity, both good and bad. It can contradict itself; thus it is one of the principles of universal explanation.[38]

Klein showed his first fire painting, *Feu de Bengale – Tableau de feu bleu d'une minute* (*Bengal Flares – One-Minute Blue Fire Painting*), at the Paris gallery of Colette Allendy in May 1957.[39] There he also displayed, amongst other things, sculptures, a tray of blue pigment powder and a completely empty room. As François Perrin has recently noted, it was at Colette Allendy's that Klein 'announced the future developments of his art, and a practice oriented towards space rather than the object'.[40] The empty room, the first 'Immaterial', was a critique, in part, of the gallery as 'white cube' – a sacrosanct space impervious to the world beyond it – a notion that would later be formulated, most notably, by the American artist and critic Brian O'Doherty (aka Patrick Ireland).[41] In April 1959, Klein took this 'Immaterial' to its next (logical) stage with his exhibition '*Le Vide*' ('The Void') at the Iris Clert Gallery in Paris, on which occasion he emptied the gallery of all its contents, except a large cabinet, and painted every surface white; the window of the gallery he painted blue, and he hung a large blue curtain in the lobby. Cocktails were served that apparently made visitors pee in blue for some days thereafter, something that had little to do with the abject as we would now understand it, but rather, I suspect, with the intimacies between the literal body and art, something that would soon be given form in Klein's *Anthropometries*, in which he would orchestrate naked women, who were covered in paint, to swirl over canvases and create traces or outlines of their bodies. There is, perhaps, an unsettling coda to Klein's '*Le Vide*': Sylvère Lotringer reminds us that the exhibition coincided with the coup enacted by rebel generals in Algiers, and Paris was awash with violence and popular protests, 'hardly a time to celebrate sensibility and nothingness'.[42]

Klein's fascination with the colour blue at Iris Clert was anticipated in two pieces at Colette Allendy's: his tray of pure blue pigment and the fire painting *Feu de Bengale – Tableau de feu bleu d'une minute*, which consisted of a plywood panel painted blue on which sixteen rockets were arranged in rows of four and pointed skyward. Placed outside the gallery, it was lit on the day of the opening.[43] Pierre Restany (the passionate champion of Klein whose many writings on the artist need to be treated with a degree of caution because of their partiality) describes how

> The rockets gave off an intense blue effect, the very image of blue fire in the chromatic definition by Gaston Bachelard that Klein obviously took note of as his motto.[44]

In so many ways Klein's muse, Bachelard understood blue to be the colour of the void; and it is this that would translate into Klein's oeuvre as 'International Klein Blue' (IKB), which he patented, and whose precise chemical formula he anxiously guarded. (Contrary to the Klein mythology, IKB was not the brainchild of the artist himself, rather, it was developed at Klein's 'request' by an employee of the chemical and pharmaceutical company Rhône-Poulenc).[45] Blue – the gateway to the ineffable, as Klein/Bachelard would understand it – and its close affinity with fire; these would become Klein's twin obsessions: 'I was immediately drawn to see the immense possibilities of this ultra-living element', Klein said of fire,[46] a phrase he took once again from Bachelard's *Psychoanalysis of Fire*: 'If all that changes slowly may be explained by life, all that changes quickly is explained by fire. Fire is the ultra-living element.'[47]

From 1957 to 1962, in collaboration with the architect Werner Ruhnau, whom he had met in Paris, Klein reflected on the Architecture of Air, a utopian and somewhat arcane project that envisioned humans living in harmony with the planet through a judicious use of air, fire and water. These meditations have largely been sidelined by Klein's paintings, sculptures and performances, but more recently they have attracted attention in their own right and were assembled by the curator, François Perrin, for the exhibition, 'Yves Klein: Air Architecture', at the MAK Centre for Art and Architecture, Los Angeles, in 2004. Speaking to Perrin about his early involvement with Klein in *Air Architecture*, Ruhnau says:

> In his work, colours were a climate of the space. What he was seeking was a paradise. He wanted to have a good climate, so we proposed to build environments that would provide this good climate. I think it was Yves who came up with this idea of the roof of air, to have a protection and air conditioning the space [sic]. I understood it instantly and said, 'Let's do some experiments!' That was in 1958.[48]

One year later, Klein would conclude his lecture at the Sorbonne by quoting directly from the Air Architecture manifesto prepared by Ruhnau and himself:

> First, blue light as a building material corresponds to void: energy must be used in the stratosphere; in the atmosphere, heavy air or any other gas that is heavier or more dense than air will be used as building material, also using *optical fire effects*, magnetism, light, and sound.[49]

Fire had been introduced earlier in the lecture and in the most excited of ways: 'Fire for me', Klein said, 'is the future without forgetting the past. It is the memory of nature.' He then went on to describe the genesis of his project for a public square with a pool where jets of fire would replace jets of water. Fire and its nemesis water, brought together:

> I was in rapture at the Granja [in 1951], a summer palace of the Spanish monarchy . . . looking at the water fountains and jets in the gardens . . . That is where I imagined replacing the tranquil surface

> of these pools, the elegant jets of water with brilliant jets of fire . . . The functional-psychological goal of water jets on stretches of water is to bring a general coolness, or, at least a sensation of coolness. For countries with a less favourable climate, where the cold reigns for quite a long time in the winter, it is a luxury to present jets of water. Whereas it is quite functional and also aesthetic and psychological to presents jets of fire on a spatial water base.[50]

These imaginings of fire would soon give way to *Fire Column* (supplied by butane gas from underground pipes) and *Fire Wall* (a grid of Bunsen burners) that Klein set up in February 1961 in the garden of the Museum Haus Lange, Krefeld, which was showing a major exhibition of his work (illus. 71). For Restany, one of the guests at the opening, 'The lighting of the flame during the vernissage had the moving character of a mystical event.'[51] Years later, the scholar Sidra Stich would write:

> One of the most spectacular parts of the Krefeld exhibition was the fire display. On the opening night, guests witnessed a dramatic inauguration when, at the appointed time . . . columns and a wall of fire suddenly illuminated the twilight . . . Even after the opening, the sensorial – and sensational – images of fires that greeted visitors . . . remained a dramatic, stirring sight.[52]

Here Klein's use of fire is testimony to its drama and intensity. And this is echoed in the 'fire paintings' he made at the Gaz de France factory, a few months later, producing some 30 fire paintings with the aid of a flame-thrower on specially prepared cardboard (illus. 72, 73, 74). For some of these works, he used a wet naked woman to make an imprint on the surface (redolent of his *Anthropometries*) before he applied the flame. In these instances, the damp areas withstood fire longer and 'the imprint', as Hannah Wietemeier has said, 'remained like a ghost shadow within the surrounding dark, scorched ground'.[53] Understandably,

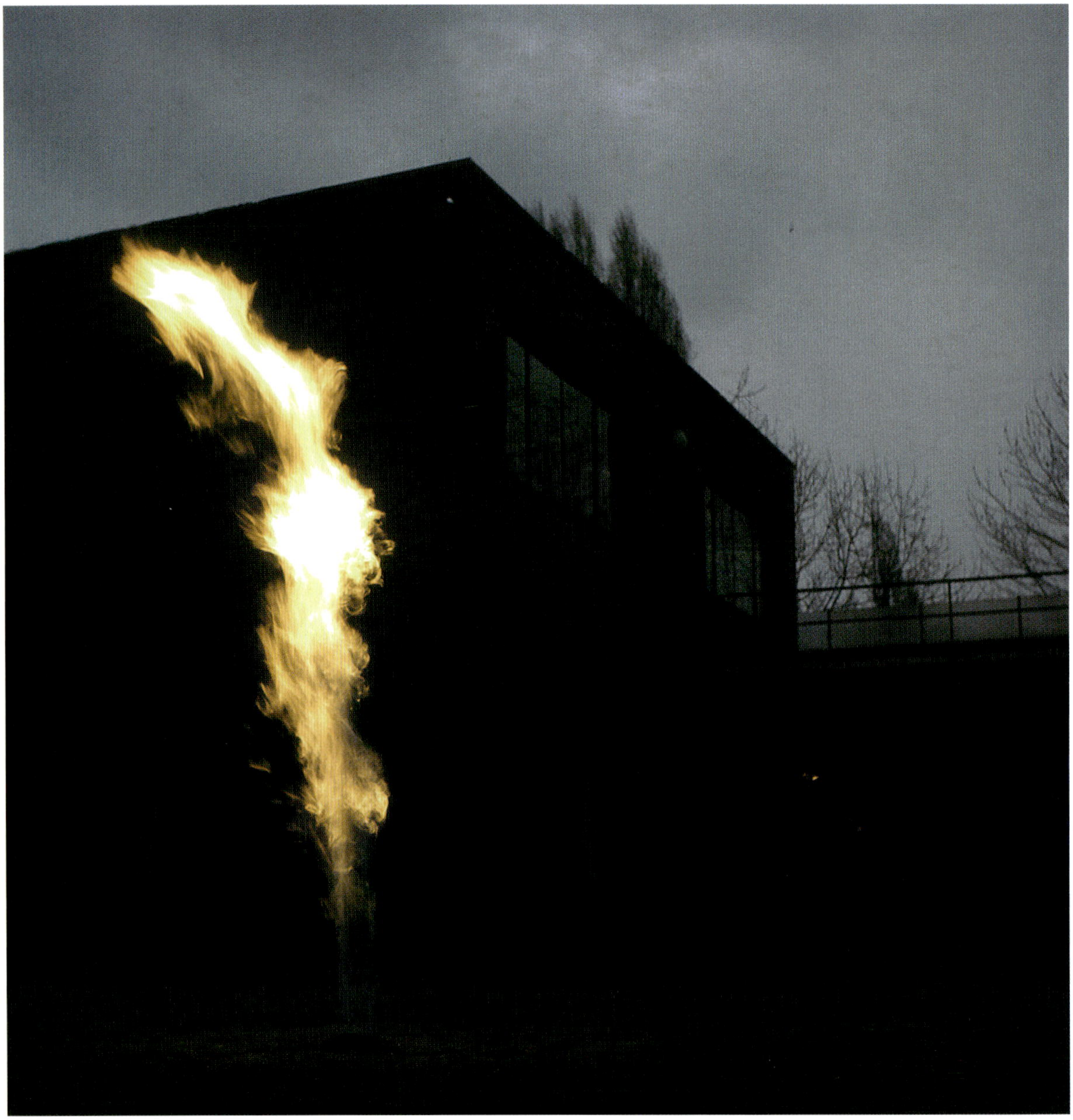

71 Yves Klein, *Fire Column*, set up in the garden of the Museum Haus Lange, Krefeld, February 1961.

commentators at the time associated such images with recent experiences of devastation and carnage; and Klein himself alluded to 'the ashes of Hiroshima blown by the wind'.[54] Yet I'm not sure that he ever took account of the socio-political *per se*. His brief, rather, was essentially utopian, transcendental and trans-national.

There's a postscript to Klein's activities at the Gaz de France: here and in other works, variously as artist, performer and always *provocateur*, Klein would make sure that all was meticulously documented. In this respect, he was one of the first to recognize the importance of photography and film as a means to nurture and market the 'image' of the artist. For his actions at the Gaz de France, Klein, as always, was well dressed, suited at times, waistcoated at others: the very embodiment of the perfect bourgeois, as Anja Zimmermann has said.[55] But he is not alone: in some of the documentation we see a 'fireman' with a water hose standing by Klein's side looking earnestly at the image unfolding on the canvas. This was no real fireman, however, but Klein's friend, Alex Costa, dressed in full gear and adding to the spectacle an extra flourish.[56] A hoax. But one on a less grand scale than the extraordinary photograph of Klein performing his 'Leap into the Void', a seeming testimony to his superhuman abilities to survive such a leap. Reproduced in various formats and contexts, this photograph 'showed' what in fact was a highly controlled event (by Klein and other helpers) followed by a masterly manipulation of the image (by its photographer, Harry Shunk) but one that carried with it an aura of credibility. 'One can turn one's nose up at Klein's clownish theatre', Yves-Alain Bois has written recently, but there is no doubting, he says, the artist's talent at challenging 'the high bourgeois culture of his time'.[57] Bois is correct.

There is 'theatre' also in the work of Ana Mendieta, but it's of an entirely different order to Klein's. Setting up *mises-en-scènes* often involving herself and then, like Klein, having them documented in photograph and film, Mendieta produced a singular body of images in the 1970s about sexuality and identity that continue to provoke and

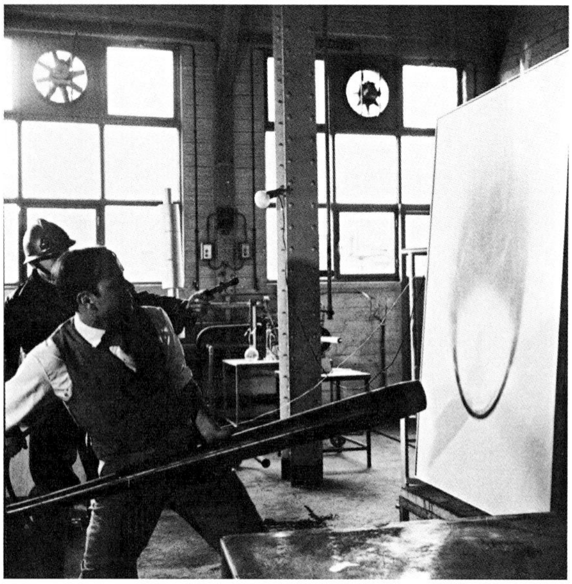

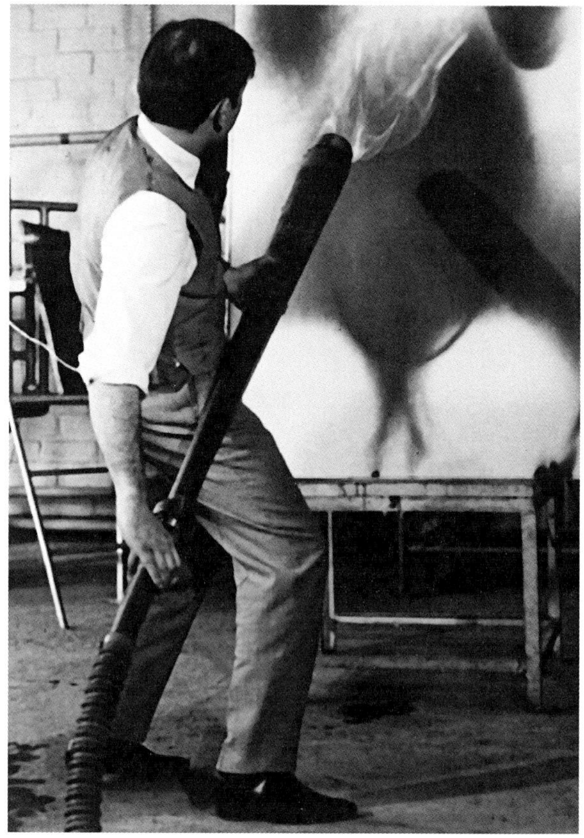

72, 73, 74 Yves Klein creating fire at Gaz de France, Saint-Denis, 1961.

resist easy classification. Like Klein, she was deeply attracted to fire's creative and transformative qualities: 'Fire has always been a magical thing for me, fusion . . . its transformation of materials.'[58] In sharp contrast to the French artist, however, she would connect her own body, literally, to the production of the unique '*Silueta*' series, which often employed fire, by introducing her naked figure (or its outline) into the landscape. (Klein as we have seen used a naked woman to help produce some of his 'fire paintings' at the Gaz de France, while he had naked women perform as 'living brushes' in his *Anthropometries*. Mendieta, on the other hand, was always the active subject and object.)[59]

In 1961, together with her sister, Raquelín, Mendieta had come to the US; her mother and brother followed them five years later. She studied at the University of Iowa, Iowa City, where she formed a professional and romantic relationship with Hans Breder, a professor of art at the University and co-founder of the Center for the New Performing Arts.

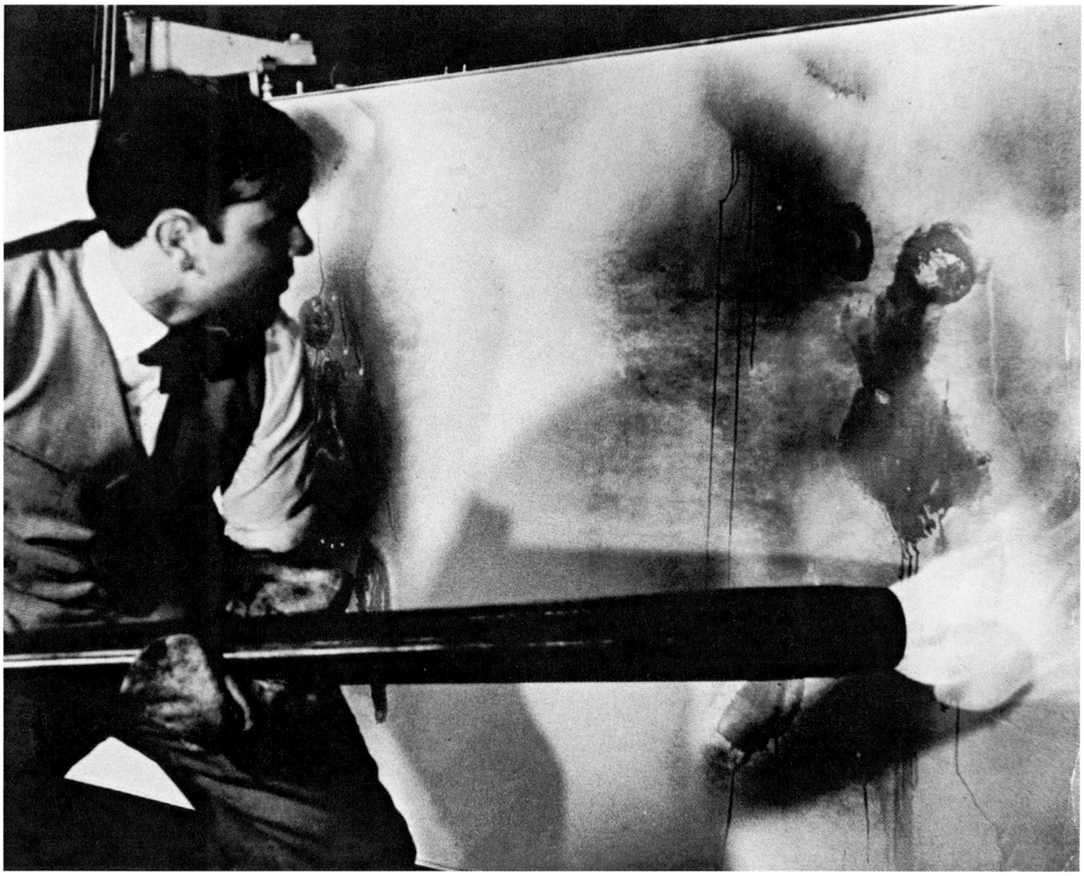

It was there too that she met another academic, Julius Schmidt, who advised her about pyrotechnics. In autumn 1975 on the banks of the Iowa River, she created her first *Silueta* with fire, *Alma silueta en fuego* (*Soul Silhouette on Fire*; illus. 75). Mendieta made it by digging a shallow outline of her form in the earth into which she inserted a life-size effigy of her body covered in fabric with arms lifted. She then set the work on fire at sunset. Olga Viso, the foremost authority on the work of Mendieta, has written that:

> The artist's use of fire was inspired by a variety of Mexican popular traditions, including the lighting of candles around graves for Nocturnal Day of the Dead vigils, the ritual burning of papier-mâché figures (Judas figures) during Easter season festivals, and the burning of elaborate constructions of cane and fireworks (*castillos*, or castles) for seasonal church activities.[60]

These instances in which fire typically plays out some of its shifting symbolic, and contradictory, roles – here it is death, betrayal, celebration – also no doubt were known to other Latin American artists who, like Mendieta, were fascinated by fire. As the writer and curator Guy Brett has suggested, these artists took 'primordial categories, such as earth or fire, and worked their cosmic nature into metaphors for contemporary social struggle'.[61] He discusses, among others, Hélio Oiticica, Cido Meireles and Leopoldo Maler, the latter of whom met Mendieta at the University of Iowa in the 1970s, and whose *Homage* (1974), a typewriter with flames coming out of its carriage, was a tribute to his uncle, an eminent left-wing journalist who was murdered in the lead up to the coup that saw the dictator Jorge Rafael Videla come to power in Argentina in 1976.

It may not be accidental that Mendieta's *Silueta de cohetes* (*Silhouette of Fireworks*) and *Burial of Ñañigo* were produced in that violent and turbulent year; 1976. Mendieta was in New York, where *Burial of Ñañigo* was first shown at 112 Greene Street, an alternative gallery: the candles outlining the silhouette of the artist burn little by little until they dissolve into wax on the floor. In the summer of 1977, she produced her first *Siluetas* with gunpowder, which she would ignite; most of these pieces were made on the property of Julius Schmidt in Iowa. Hitherto unseen colour slides documenting these works (and others) have only recently been published: they show wispy smoke rising from *Siluetas* made in green fields or in what looks like dry, cracked river beds: at times, the white outline of the body is still sharp, at others, all that remains are black ashes with tiny pockets of red flame.[62] Plaintive though these images are, there is a hint of redemption, as it were, in the surrounding, verdant grasses. A remarkable series of slides from an untitled gunpowder work of 1981 show the white silhouette located, incongruously, in dunes (Indiana Dunes National Lakeshore).[63] Flames and smoke collapse into a frothy concoction of orange, brown, yellow and silver-grey; and all this is set against the mushroom colour of the sand, which in these close-up

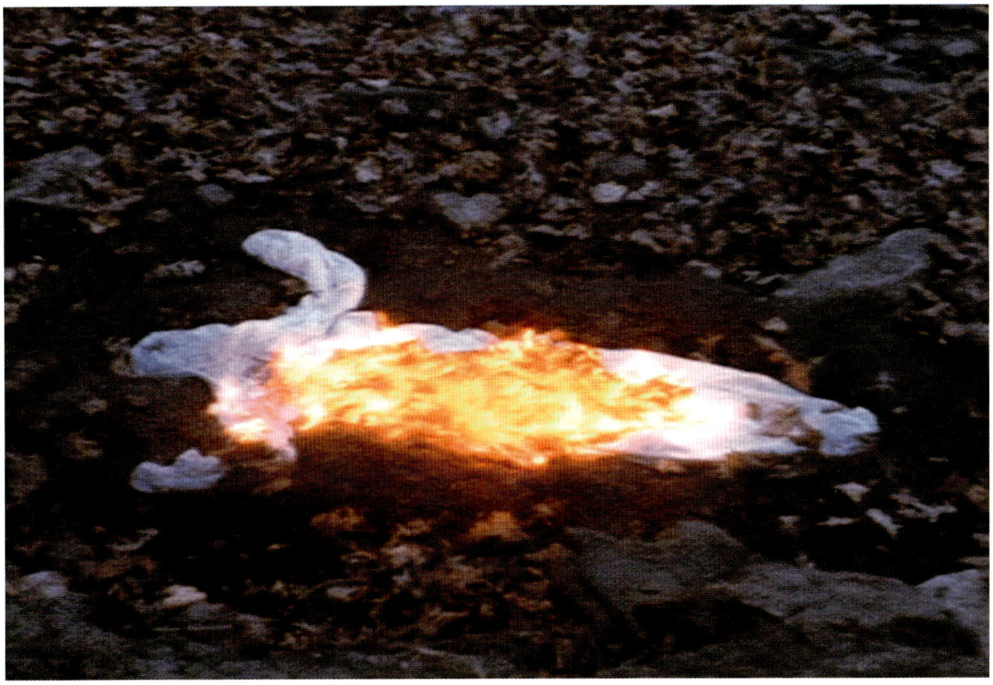

75 Ana Mendieta, *Alma silueta en fuego* ('Soul Silhouette on Fire'), 1975 (Iowa). Lifetime colour photograph from 35 mm colour slide.

shots reveals specks of detritus and miniscule craters. At the very top of the images are indistinct, blackened trees; these are more conspicuous in one slide, where their (seemingly) wintry silhouettes (the trees are denuded of leaves), make one see the sand, suddenly, as snow! This double take is set against the flames arising from the whitish shape of the *Silueta*, flames so feathery and soft, one almost wishes to caress them.[64] Isolated on a page of a book – the way in which I've seen the image – this particular slide has about it the quality of a magical, painterly conceit. By contrast, two slides documenting an untitled piece actually made in the snow at Iowa in 1977 show the *Silueta* photographed, unusually for Mendieta, from almost directly above, leaving the viewer looking down at a deep black, shroud-like *Silueta* with interruptions of flames: it lies in a field of snow, a silent, haunting presence.[65]

Mendieta employs fire in ways that embrace its dualities: death and rebirth, purification and memorializing, the private and the public. Christian and pagan sometimes rub shoulders; at others, they retain their distinctiveness. Her last gunpowder *Silueta, Mayora* (1982), commissioned by collectors in South Carolina and made at the back of their home, was named after an Amerindian moon goddess. Constructed of concrete and earth, it testifies, as Olga Viso has said, to Mendieta's focus in the early 1980s on outdoor works of a less ephemeral nature. The piece was lit at night before a small gathering of friends and supporters.[66] Left behind was a molten form set in the earth, which Mendieta encouraged the owners to relight on special occasions in the future.

Ana Mendieta's lifelong fascination with the body and fire contrasts with the restricted and quite different but memorable uses to which Gina Pane puts it. Of Italian origin, Pane was born in Biarritz in 1938 and died in Paris in 1990. A radical figure in the performance art of the 1970s, Pane 'pushed herself to her physical and psychological limits'[67] in an action titled *Escalade non Anesthésiée* (*Unanaesthetized Climb*), performed in her studio in Paris in 1971, the artist repeatedly climbed a metal ladder with sharp points on the rungs (her intention being, as she later said, to counter apathy in the face of events such as the Vietnam War).[68] On other occasions, she would cut her skin with a razor blade. These physically demanding performances paralleled other performance artists such as Dennis Oppenheim having rocks thrown at him from above *(Rocked Circle – Fear*, 1971); Linda Montano placing acupuncture needles around her eyes (*Mitchell's Death*, 1978); Marina Abramović's *Rhythm O*, performed only once in Naples in 1974, in which she invited members of the audience to do anything to her body with items such as knives and a loaded gun, and Chris Burden in *Prelude to 220, or 110* (1971), who had his arms and body tied and bolted to the gallery floor with copper bands. Nearby were placed two buckets of water with live 110-volt lines in them. As Cynthia Carr has written pointedly,

> had any visitor chosen to spill the water, Burden would have been electrocuted. Typically, he was forcing himself, the audience, and the sponsoring institution to face an elemental and harrowing reality.[69]

These artists made risk and injury and ultimately death, at the most extreme, a model for their practice: in part a reaction to the 1960s, one suspects – not its 'sex and drugs and rock and roll' narrative that had, of course, its own victims, but rather to Vietnam, political assassinations, the traumas of the Civil Rights Movement and the workers' and student protests of May 1968 in Paris.

Gina Pane, it would seem, used fire on only two occasions: the first in 1971 during a performance called *Nourriture, actualités télévisées, feu* (*Food, TV News, Fire*) held in a private apartment in Paris, and the second in a three-part action, *Autoportrait(s)* (*Self-Portrait[s]*), performed at the Stadler Gallery in Paris two years later. In *Nourriture*, she commenced by eating raw ground meat, which she then spat out. The next sequence consisted of her sitting with other participants and watching the evening news (coverage of the Vietnam War), all the while having the bright light of a lamp glaring directly into her eyes. Finally, she tried to put out a small fire lit on a pile of sand using her bare hands and feet. Engagement versus indifference, ingestion (in all senses) and displeasure, vulnerability and suffering: these constitute the central themes of this performance. Fire is not employed as an element of cleansing and transformation, as it was with Mendieta, but simply as an element capable of producing pain. *Nourriture*'s direct referencing of the Vietnam War and its media coverage now take on added meaning. The date and year when the action took place, 24 November 1971, is often overlooked (perhaps it is not even known by some scholars) but it is significant.[70] The early months of 1971 saw the launching of Operation Lam Son 719, a massive attack by South Vietnam on North Vietnamese troops in Laos. On both sides, but especially in North Vietnam, there were huge fatalities, some 20,000, while South Vietnam lost more than 7,600 and the US over 200. Then, on

13 June of the same year, the *New York Times* published parts of the 'Pentagon Papers' (released to the newspaper by Daniel Elsberg), that infamous document detailing the deceit of various US administrations over Vietnam. All this would not have been lost on Pane, who would go on to say about *Nourriture*:

> It's strange, we never felt or heard the news before. There's actually a war going on in Vietnam . . . Until this moment, they were anaesthetized in the face of world news.[71]

In *la mise en condition* (*The Conditioning*), the first part of *Autoportrait(s),* Pane lies fully dressed in dark blue trousers and a light blue, open-necked shirt, on an iron, bed-like structure, beneath which fifteen candles burn, their flames rising to within inches of her back (illus. 76). (The pioneering performance artist Marina Abramović provocatively re-enacted this action and six other seminal performances from the 1960s and '70s in an exhibition titled 'Seven Easy Pieces' at the Guggenheim Museum in November 2005.)[72] Lasting 30 minutes, Pane's *la mise en condition* involved her clasping and twisting her hands. She would later say:

> The public understood my suffering from the way I wrang my hands much more than from my face [the documentation shows her looking serene], so it was actually a very primitive mode of communication. But I feel I succeeded in making the public understand right off that my body is my artistic material. When, half an hour later, I was able to get up, I caressed my body very gently. There was no violence; my body hurt but I could still feel my touch.[73]

The Australian Mike Parr (*b.* 1946) has, like Pane, pushed his body to its physical and psychological limits. Of performances during

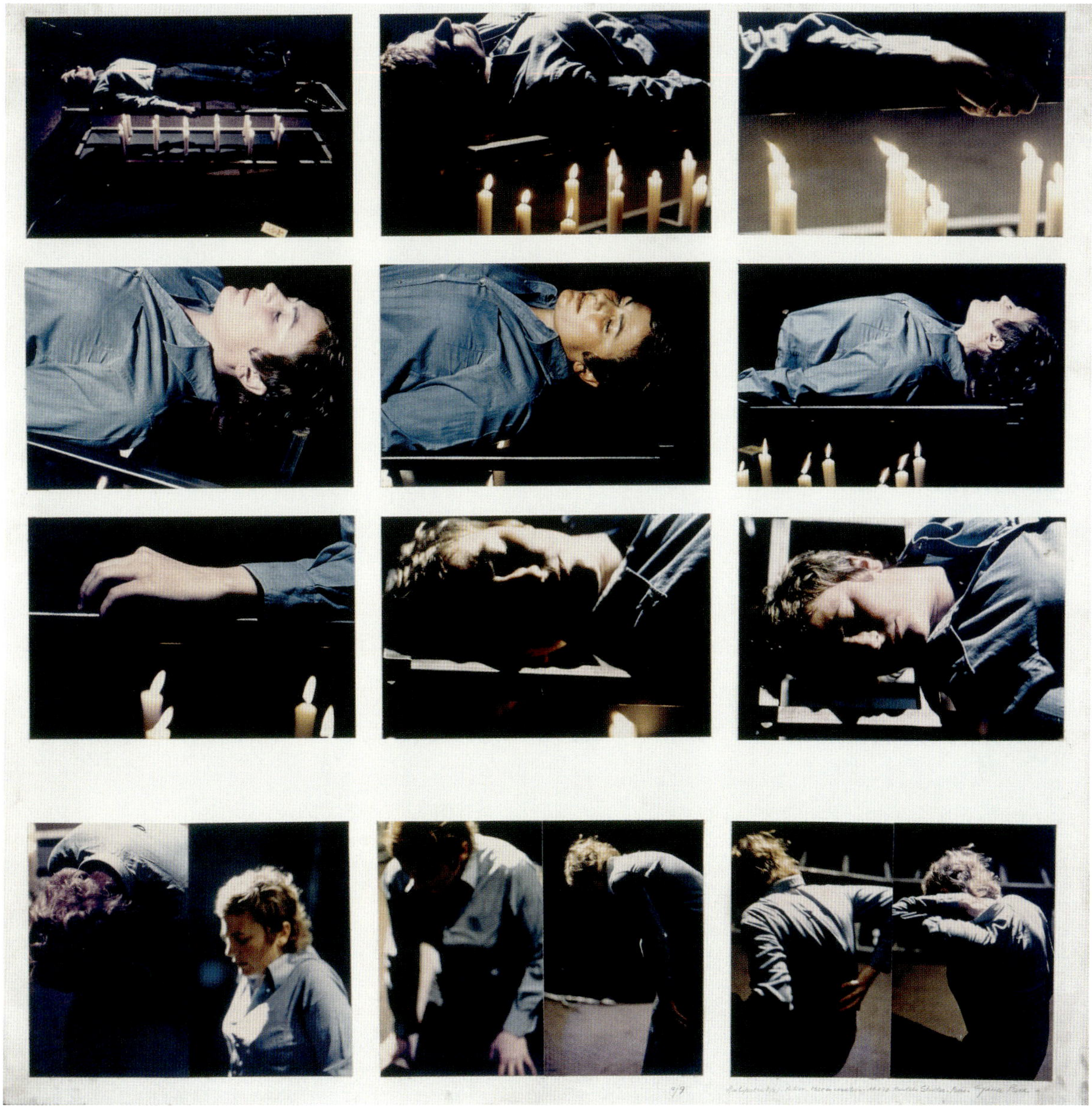

76 Gina Pane, *Autoportrait(s): la mise en condition, la contraction, le rejet* (1973).

the years 1970–74, the period in which he used fire most prolifically in his work, he wrote at the time about his increasing interest

> in the relationship between extreme instruction pieces (*Drop a brick on your foot*) and the actual implication for the instruction of carrying them out . . . I had also begun to understand the implications for art as a whole, of the idea of art as a means of behaviour modification. I began to see that the performance by the artist of simple, meaningless acts of violence towards his own person, in the name of art, placed the whole art paradigm, including its audience, in a tremendous quandary. How was such an act to be assessed? How could responsibility be attributed (if the work was performed before an audience) or accepted? What about the problem of enforceable restraint and artistic freedom. Etcetera, etcetera.[74]

At Inhibodress Gallery, Sydney, in 1972 (an artist-run cooperative he had set up with a close colleague, Peter Kennedy), Parr preformed *Hold your breath for as long as possible* and *Light a candle. Hold your finger in the flame for as long as possible* (illus. 77); both these were filmed in private. In the former, we see Parr's head jerking from side to side, eyes rolling, until finally he takes a huge, shuddering breath. The unedited black and white film of *Light a candle* shows in close-up the burning of Parr's finger: we hear the sounds of this burning, the artist's cries of pain and finally his screams. Here fire (or flame, if you will) is framed as simple, brutal fact. Also at Inhibodress in 1972, but this time in front of a live audience, Parr performed his controversial *Setting before an audience . . . bare your shoulder. Let a friend bite into your shoulder . . . until blood appears*. Reviewing the action, the art historian and critic Terry Smith articulated forthrightly his ambiguous impressions:

> Peter Kennedy bit into Mike Parr's upper arm as hard as he could for three minutes or so. No blood was drawn. Parr's body shook

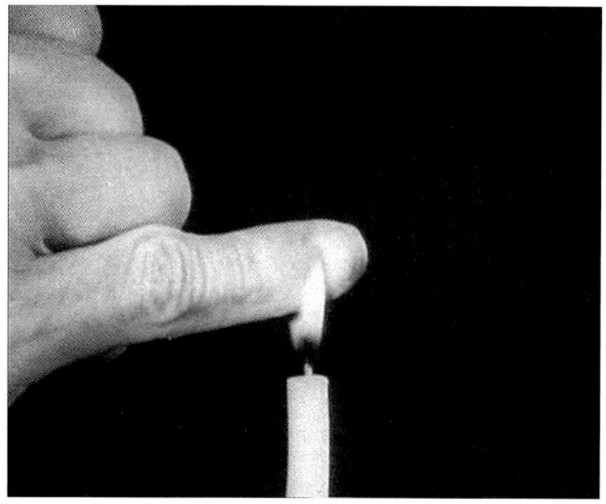

77 Mike Parr, *Light a candle. Hold your finger in the flame for as long as possible*, 1972.

> with great pain, and the performance stopped only when Kennedy could no longer continue biting. Try as I might, I could not discover in myself a clear response towards this, of either repulsion or acceptance.
>
> Parr did make every effort to cancel any masochistic intent the work might have or convey [he in fact made a formal statement at the beginning of the piece disavowing any sadistic or masochistic intentions]. There was no sense of invitation to the audience to participate or even sympathise with the work.[75]

After the action Parr engaged his audience in debate. According to David Bromfield, the author of a major monograph on Parr, the artist says he recalls Terry Smith's response as 'very agitated'. He kept saying: 'Why did we let this happen?'[76] This question of course drew attention to the audience and its ethical role, something that would resurface in the action *Have a burning match dropped on your bare chest (Wound by Measurement 4)*, performed both in Lausanne, at Galerie Impact, and in Sydney at Central Street Gallery (illus. 78). In a letter to Rosemary Schwarzwälder, June 1973, Parr wrote tellingly about the Lausanne performance:

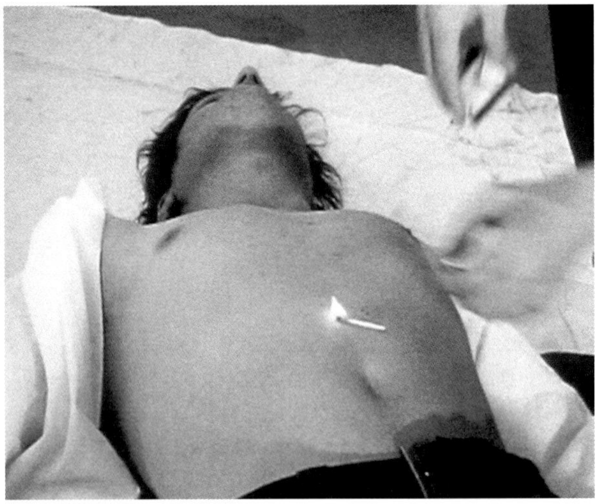

78 Mike Parr, *Have a burning match dropped on your bare chest (Wound by Measurement 4)*, 1973. From the series: *Rules and Displacement Activities Part 1*, May 1973.

> In Switzerland I asked members of the audience to drop burning matches on my body and to continue to do so until I asked them to stop. I didn't ask them to stop and the piece proceeded until a) they decided to stop individually or as a group, or b) I started to get burnt and discontinued . . . I can't quite remember now. I can remember telling them to keep on going when they had begun to waver . . . But anyway, the objective was to get this Swiss audience to say 'non' – to stop thinking of the piece as 'art' and start raising moral objections.[77]

In May 1973, also at Galerie Impact, Parr did another action involving matches, titled *Thresholds No. 1; With Fire and Water*. Bringing together these intrinsic opposites, the piece involved participants cupping their hands into which water was poured; Parr then burnt matches in front of their faces (their eyes were closed), until 'they either sensed the light or felt the heat'; finally, he had the participants look at their hands for traces of water.[78] The photographs documenting this performance show participants with their eyes shut, smiling and (seemingly) content: they could be involved in some deeply moving and pleasing religious ritual. This is not the flame of aggression or pain evinced in Parr's previous performances, but rather, one that is gently redemptive.

This quality is also suggested in the photograph of the Polish artist and activist Jerzy Bereś (*b.* 1930) taken during a performance at the Krzysztofory Gallery in Kraków in October 1979 (illus. 79). Trained in Kraków at the Academy of Fine Arts, Bereś went on to become a central figure in the dissident art of the 1970s and '80s in east-central Europe. He was, however, largely overlooked in the West at the time, with the possible exception of the extraordinary documentary directed by Ken McMullen and written by the seminal British performance artist Stuart Brisley, *Being and Doing* (1984), which shows, amongst other examples of performance art, Bereś's *Romantic Manifestation* of 1981, to which I will return.[79]

Bereś frequently used fire as a vehicle in his actions. As early as 1968, and largely in response to the Communist regime's severe censorship of the press during a time of ruthless anti-Semitism, he staged *Prophesy II* at the Krzysztofory Gallery, during which he and members of the audience burnt copies of the Warsaw-based weekly *Kultura* (*Culture*). At the conclusion of the action (all the while Bereś was draped in white and red cloth, the colours of the Polish flag), he asked for a piece of burning wood, blew out the flames and signed his 'work' with the charred stick.[80] Here 'creativity', it seems to me, is linked unmistakably to fire's Promethean capacities.

In the work reproduced here (see illus. 79), Bereś, then almost 50 years of age, stands naked before a pile of burning logs. (He always appeared nude in his performances, something that Piotr Piotrowski has written suggestively about in his recent study of avant-garde art in Eastern Europe.)[81] Jacek Szmuc, who now lives in Melbourne, Australia, and with whom I have had the good fortune to speak, took the photograph. What follows are his translations of the words that Bereś used in the action, their nuanced meanings, especially for a Polish audience.[82] The word WARSZAT on the central, circular piece of wood ranges in meaning from 'workshop' to 'worktable' to the way 'one does things'. CZYSTE (on the mat) evokes the 'pure, the clean, the clear', while beneath it, DZIEŁO, might be 'a work, a piece of art, a done thing, an achievement': 'Dzieło Sztuki' means a work of art. Finally – perhaps we should have begun with this – there are the three letters 'AKT' emblazoned on Bereś's torso. These, Szmuc says, could stand for 'An act (as in act or gesture) – an action', but also, 'an arty picture of a nude would be called AKT'.[83]

This subtle shifting and collapsing of linguistic signs mix the aesthetic and the political. The performance took place in late November 1979; less than a year later, in Gdańsk in September 1980, Lech Wałęsa would co-find Solidarity, the first non-communist party-controlled trade union in a Warsaw Pact country. Bereś described the Solidarity movement as 'a creative fact', a term, as Klara Kemp-Welch has pointed out, 'he generally used in place of the word "art"'. And she continues:

> The success of a creative fact was for Bereś measured by its capacity to interrupt the suspension of judgment that all too often is at the root of social as well as aesthetic stagnation . . . His manifestations from this period were tied to the struggle to provoke judgement.[84]

In his famous *Romantic Manifestation* performed in the main square of Kraków on 18 November 1981, one month before Martial Law was declared, Bereś turned to fire again. As mentioned earlier, this action is shown in the documentary, *Being and Doing*. The voiceover describes the event:

> I saw Bereś perform in the city square . . . People were queuing for food at all hours . . . [He] pushed a wooden handcart he had made into the square. He had a slab of wood sticking up from it on which were fixed bundles of kindling sticks, each one labelled with a word. He wrote one of the five words, 'freedom', in a circle on the ground.

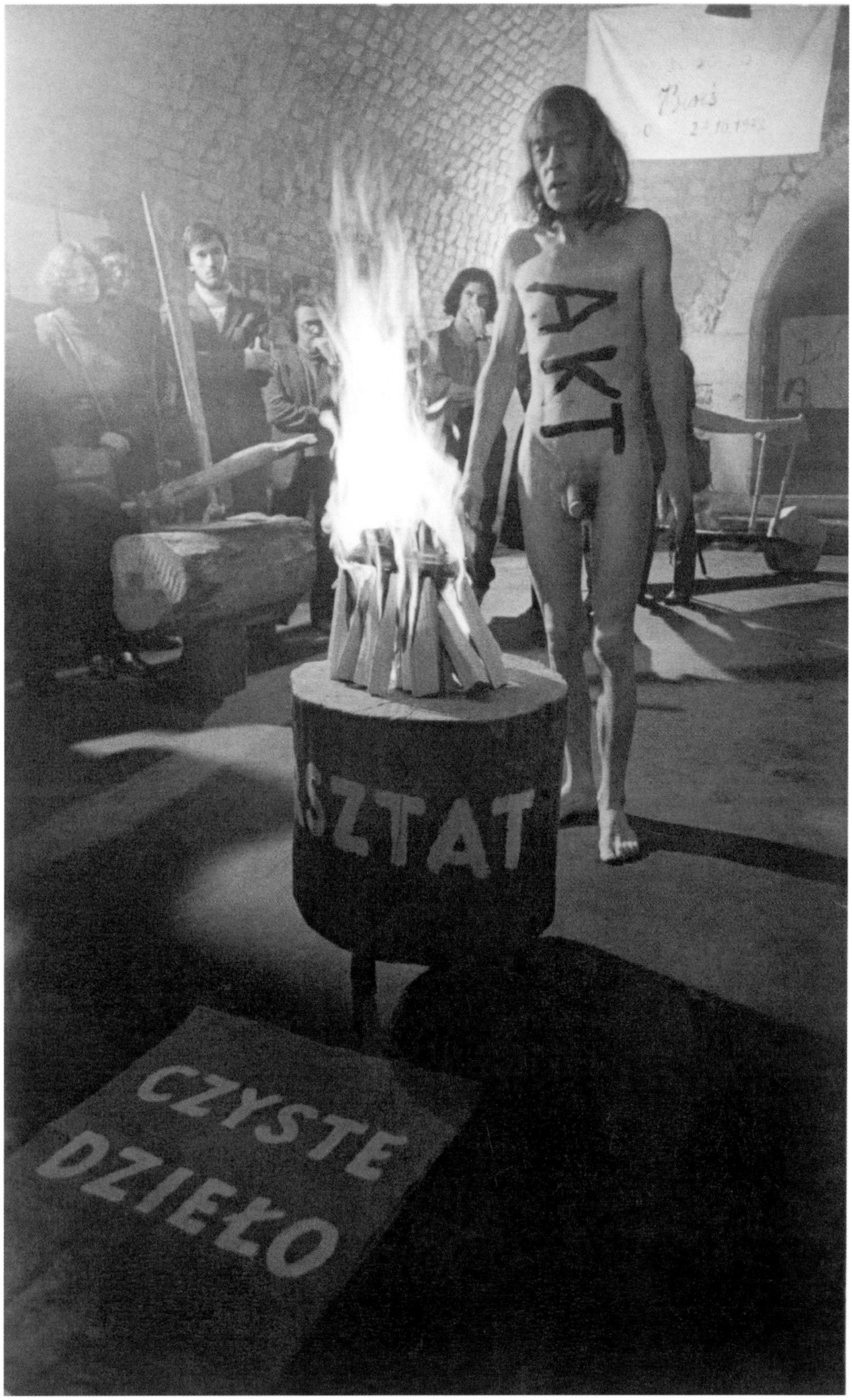

79 Jerzy Berés, Performance at the Krzysztofory Gallery, Kraków, October 1979.

> And then he took one of the bundles of sticks and lit a fire in the circle. And as he slowly repeated the action four times, hundreds of people gathered, they were tightly packed, leaving small clearings around the fires, each with a small wisp of smoke rising into the air.

The words that Bereś used to describe the small fires were in fact: 'fire of hope', 'fire of dignity', 'fire of love', 'fire of truth' and 'fire of freedom'. Here the overlapping of the religious and the secular/political is obvious. Not surprisingly, these words had been taken away by the morning after the performance. What was surprising for the artist, however, was that the same thing happened when he did the action again in November 2000, at a time when the first post-communist Constitution of Poland had been in place for over three years.[85]

The Hearth Revisited

In 1969, on every night between Christmas Day and New Year's Eve, the German TV channel WDR broadcast the Dutch artist Jan Dibbets's (*b.* 1941) *TV as a Fireplace* in colour at the close of its programmes. Showing the beginning and the end of a filmed hearth fire (each segment lasted about two and a half minutes), it never mentioned Dibbets's name, nor that the broadcast was in fact an artwork. One amongst other bold initiatives of Werner Höfer, manager of WDR, the piece seemingly effaced the boundaries between art and life (a mantra of the 1960s avant-garde); to be more precise, 'life' as mediated by television. Still, it no doubt allowed many thousands of viewers, one assumes, to partake, individually and collectively, in the pleasurable experience of the (virtual) fire.[86] This is fire at its prankish best: 'It wasn't intimidating . . . it's a very friendly way of misleading people', Dibbets said in one of his very rare published comments on this work.[87]

Some fourteen years later, the American-born, London-based artist Susan Hiller (*b.* 1940), whose work traverses installation, sound, photography and investigations into the worlds of dreams and near-death experiences, amongst other things, made *Belshazzar's Feast/The Writing on Your Wall*. Central to its meaning is fire and its relationship to the hearth and to reverie, a theme that we have encountered before in the writings of Gaston Bachelard. One version of Hiller's *Belshazzar's Feast* (exhibited at the ICA, London, 1986) has four television monitors on the floor, back to back, in a configuration that suggests a campfire. Another, the so-called 'living room' version, shows a single television monitor in a meticulously arranged 'modern' domestic setting with low-lit table lamps (illus. 80). Screened on the monitor/s in all adaptations of *Belshazzar's Feast* is a compilation of constantly changing images of fire.

The segmented soundtrack to this work contains Hiller's own voice and that of a child (her son). It's a haunting mix of 'nonsense' utterances, whispered references to people who apparently had received transmissions from 'aliens' through TV after closedown, and the child's efforts to describe from memory Rembrandt's painting on the subject of *Belshazzar's Feast* in the National Gallery, London, which he had seen in reproduction. This outpouring of words cuts to the quick of the biblical account, in which a hand suddenly appears in the middle of a banquet arranged by the king and writes mysterious words on the wall. Says Hiller:

> As I understand the story, he [Belshazzar] in fact could not r*ead* the words in a literal sense but was able to *interpret* them . . . I am very intrigued by the distinction between literally reading and understanding signs or marks, and interpreting them.[88]

Reflecting, interpreting, and dreaming: all this takes place in front of Hiller's fire, the key to *Belshazzar's Feast*:

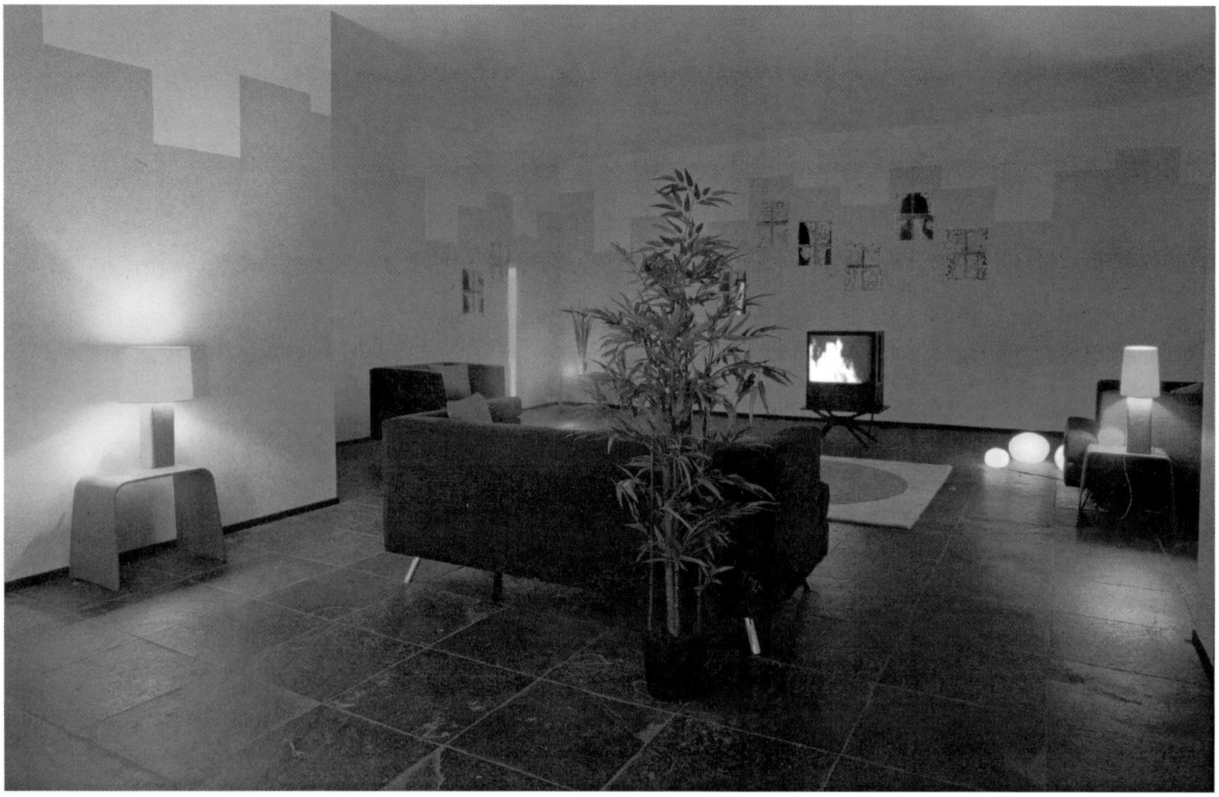

80 Susan Hiller, *Belshazzar's Feast / The Writing on Your Wall*, 1983/4, mixed media.

I have the feeling that nowadays the television set functions like the hearth or fireplace used to. What I wanted to do with the imagery of this work was to establish for the audience their own ability to dream, to create images out of these free-floating, ever-changing forms.[89]

Postscript: 'A Dreadful Prankster'

It has not been pretty, this century. And fire is everywhere implicated. Two days after the exhibition, *Burn: Artists Play With Fire* concluded at the Columbia Museum of Art, South Carolina, on 7 September 2001, agents of Al Qaeda flew two hijacked commercial aircraft into the twin towers of the World Trade Center: massive explosions of fire and falling debris and body parts greeted the still-new century. Over 2,750 lives were lost.

In actuality and its representation – in newspapers, TV and the *vox populi* world of YouTube – fire continues to assert itself; the list is long, but a few examples stand out: the 2003 invasion of Iraq; the 2005 'second' Lebanon War; sporadic burnings in South African black townships redolent of the Apartheid era (notably in Khutsong, 67 km (42 miles) outside of Johannesburg, in 2006);[1] and ongoing acts of suicide extremism. Destruction, death and fire complement each other and play out their dreadful roles. As I write (May 2010), a photograph of an Athens riot police officer felled by a Molotov cocktail (he struggles behind a wall of flames) is everywhere in the daily press: the Greek government's introduction of stringent measures to counteract the deep economic malaise into which the country has fallen is being met with widespread popular opposition. And in the last few days, the conflict in Thailand has escalated in remarkable ways, with the army and the Red Shirts, a loose coalition of anti-government forces, at a stand-off: images of burning tyres, buildings and vehicles are commonplace.

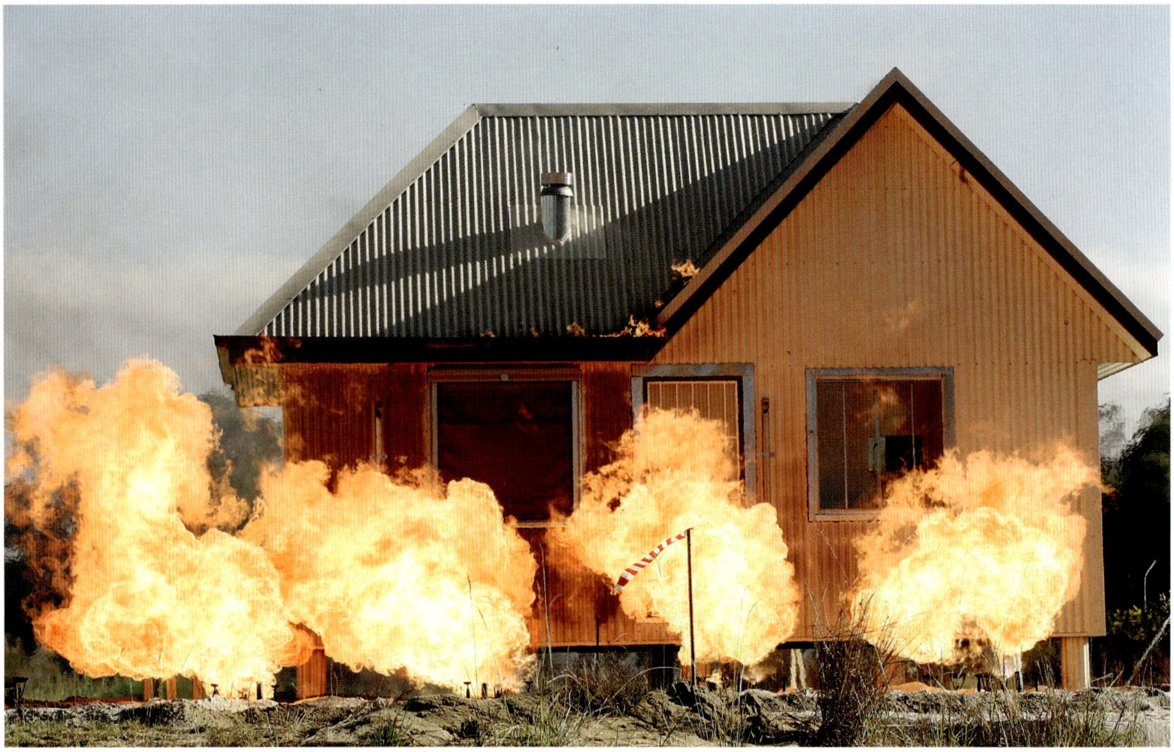

81 Before the controlled burn. *Sydney Morning Herald*, Weekend Edition, 17–18 April 2010.

All this illustrates one side of the Promethean myth. Not the wondrous creation of man from clay and water to be followed by the theft of fire from heaven to make life more tolerable, a beneficent action, but rather, the insidious, where the omnipotent god Zeus, recognizing that his temporal powers would be compromised by the advent of humans, unleashes Pandora and all hell breaks loose.

And then, of course, there is that other threat: global warming which, contrary to a small group of voluble sceptics, advances inexorably. Destructive wildfires are a common enough feature in Australia, that most flammable of all countries, but warnings have been raised more recently that such fires were 'virtually certain to increase in intensity and frequency' because of progressively warming temperatures over the next few decades.[2] According to Australia's Commonwealth Scientific and Industrial Research Organization (CSIRO), by 2020 there could be up to 65 per cent more extreme fire-danger days compared with 1990, and in the most severe scenario, there could be a 300 per cent increase.[3]

Against this alarming backdrop and the inertia of governments to reduce global carbon emissions, it should come as no surprise that other alternatives are being considered. In Australia, the National Association of Steel Framed Houses, along with the CSIRO, are working on a house made almost entirely of steel that would survive a bushfire and protect its inhabitants should it prove impossible for them to leave their property. The three photographs reproduced here (illus. 81, 82, 83) show the

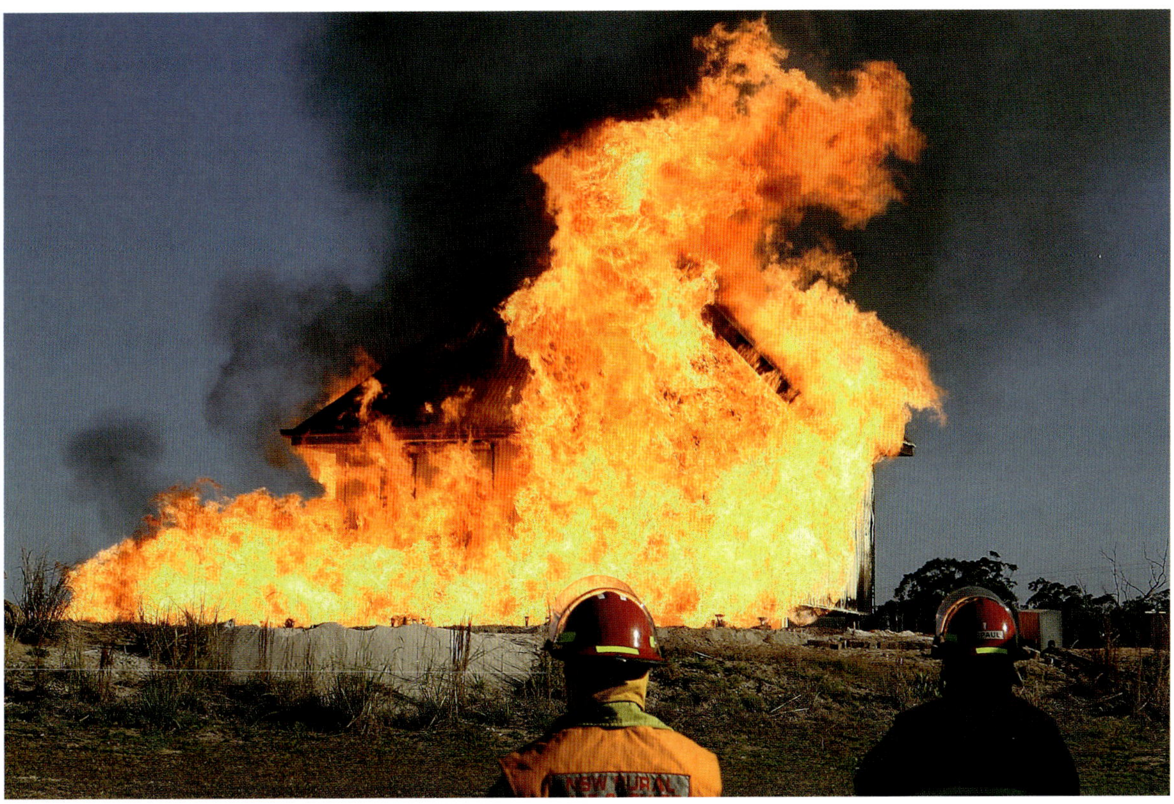

82 The controlled burn. *Sydney Morning Herald*, Weekend Edition, 17–18 April 2010.

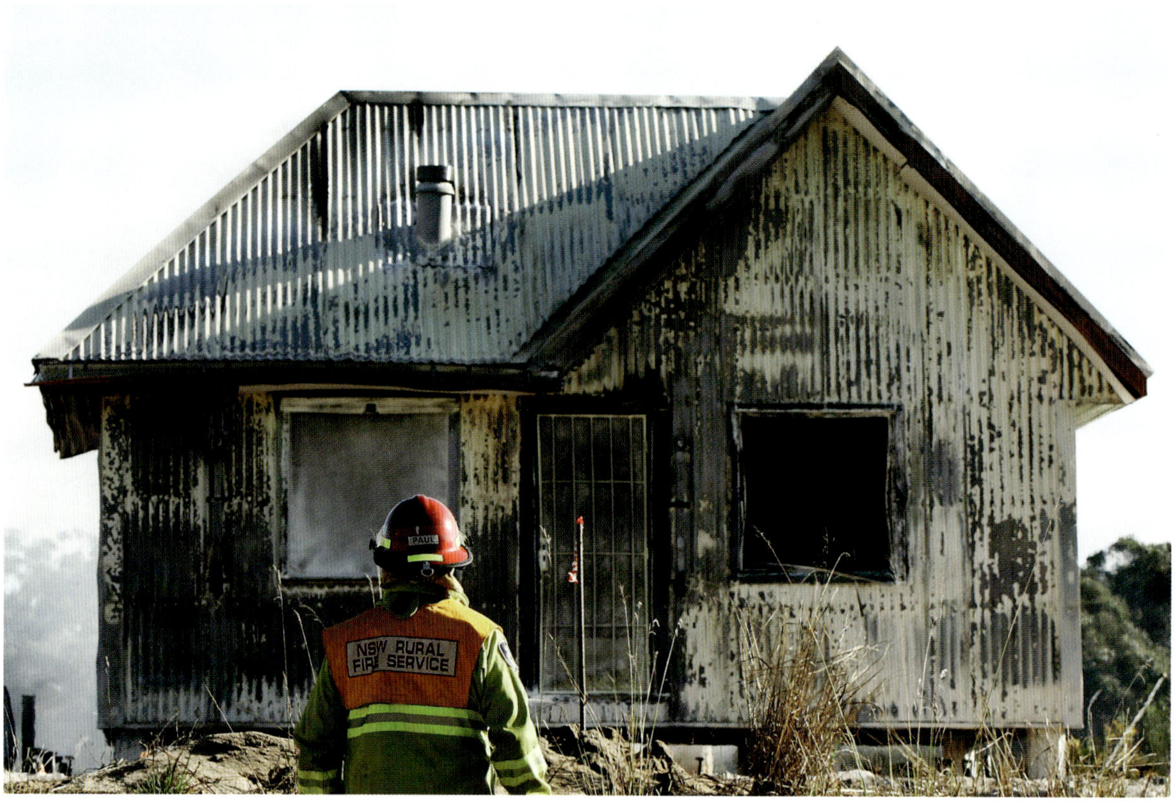

83 After the controlled burn. *Sydney Morning Herald*, Weekend Edition, 17–18 April 2010.

house before, during and after the controlled burn undertaken by the CSIRO scientist, and project coordinator, Justin Leonard, assisted by the Rural Fire Service at Mogo on the south-east coast of New South Wales. The building was first subject to radiant heat from gas burners for over 30 minutes. The intensity of the burners grew until the house was taken over by flames 3 m (10 ft) high for nearly two minutes. 'At its peak', wrote the journalist Nicky Philips, 'the temperature of the fire would have been more than 1000 degrees'.[4]

Will this new habitation allow its occupants to walk out alive and well? Well, that remains to be seen. Prometheus, after all, was chained to a rock in the Caucasus where each day an eagle bit at his liver but it was

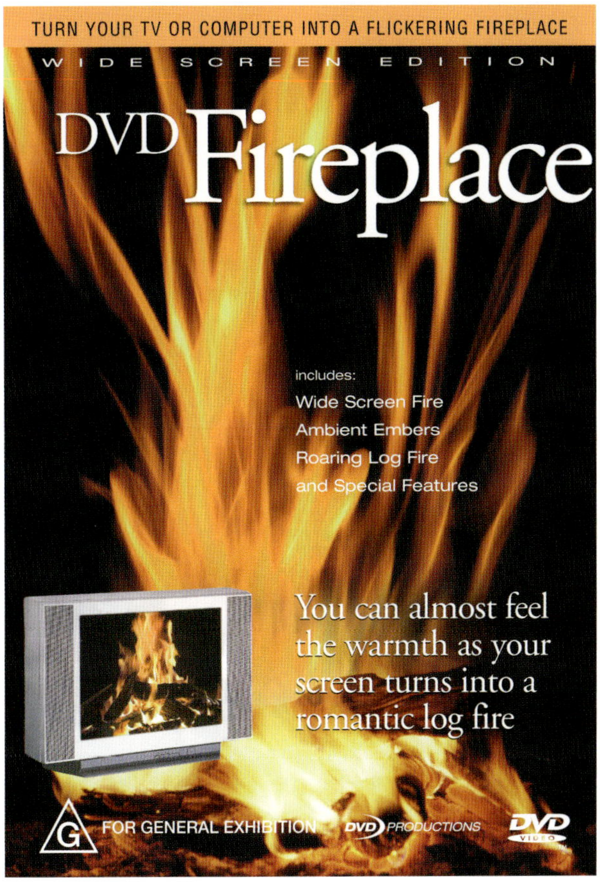

84 Mark Denning, DVD *Fireplace*, 2001, cover.

restored again at night. And the three young men, Shadrach, Meshach and Abed-nego, who refused to worship before the image of gold made by Nebuchadnezzar, the King of Babylon, miraculously survived the fire into which they were thrown.

If a house at the heart of a rampaging bushfire might survive together with its inhabitants, it is not through divine intervention but rather human agency and ingenuity. Here culture wins out over nature. In the case of DVD *Fireplace*, however, culture simply obliterates nature (illus. 84). With this device your TV or computer can be transformed into a fireplace. Continuing in the tradition of the artists Susan Hiller and

Jan Dibbets, albeit made for very different reasons and intended for a very different constituency, DVD *Fireplace* (2001), directed by the Australian Mark Denning, allows you to select either a 'Roaring Log Fire' or 'Ambient Embers' with a choice of natural sounds or piano melodies.[5] The visual delights of the hearth are there in abundance, and even its seductive sounds, but like the Burning Bush, which burnt with fire but was not consumed, this fire is definitely not for burning.

REFERENCES

Preface: *Framing Fire*

1 'California Pins Hope on Damp Ocean Breezes', *The Eagle-Tribune* (28 October 2003): www.ecnnews.com/cgi-bin/05/etarchivestory.pl?200310+fn (accessed 22 September 2004).
2 'The Expulsion from Paradise', (circa 1360–70): www.getty.edu/art/collections/objects/o2892.html (accessed 2 September 2004).
3 William Sansom, 'The Naked Flame', in *Fire and Water: An Anthology by Members of the NFS* [1942], ed. H. S. Ingham (Exeter, 1992), p. 217.
4 Ibid., pp. 217–18.
5 Ibid., p. 218.
6 Ibid.
7 Ibid., p. 219.
8 'California Pins Hope on Damp Ocean Breezes'.
9 Rupert Martin, *The Sculpted Forest: Sculptures in the Forest of Dean* (Bristol, 1990), p. 57.
10 See Stephen Porter, *The Great Fire of London* (London, 1996).
11 See Adrian Tinniswood, *By Permission of Heaven: The Story of the Great Fire of London* (London, 2003).

1 Fabled Fires

1 Reproduced in *The Great Fire: Leading Newspaper Accounts of the Terrible Chicago Conflagration* (St Louis, MO, 1871), p. 1. Document kept in the Chicago Historical Society. See Lisa Krissoff Boehm, *Popular Culture and the Enduring Myth of Chicago, 1871–1968* (London, 2004).
2 See, for example, Peter de Bolla, *The Discourse of the Sublime: Readings in History, Aesthetics and the Subject* (London, 1989) and Andrew Wilton, *Turner and the Sublime* (London, 1980).
3 At www.fire-extinguisher101.com/biggest-fires-htr (accessed 20 June 2009).
4 'The Monster Arrives Early', *Sydney Morning Herald*, 9 October 2002, p. 1.
5 *Daily Telegraph*, 9 February 2009, p. 1.
6 See State Library of Victoria, 'Bushfires in Victoria', http://guides.slv.vic.gov.au/content.php?pid=36394&sid=267991 (accessed 20 December 2009).

7 See www.bushfire.nsw.gov.au/desp_content.cfm?cat_id=1180 (accessed 1 August 2009).
8 Robert Wainwright, 'More Spin than a Warnie Wrong "un"', *Sydney Morning Herald*, 18–19 July 2009, p. 3.
9 Correspondence with the author (1 September 2006).
10 See David Buckman, 'Obituary: Paul Dessau', *The Independent*, 11 November 1999, p. 7.
11 Ibid.
12 William Sansom, James Gordon et al., *Jim Brady: The Story of Britain's Firemen* (London, 1943), n.p. See also Angela Wright, *Leonard Rosoman: A War Retrospective, 1939–45*, exh. cat., Imperial War Museum, London, and the Fine Arts Society, Edinburgh (1989), n.p., and Sally Holloway, *Courage High! A History of Firefighting in London* (London, 1992).
13 Martin Gilbert, *Second World War* [1989] (London, 2000), p. 123.
14 Bernard Reilly Jr, *Currier & Ives, A Catalogue Raisonné*, 2 vols (Detroit, MI, 1984). See also Harry T. Peters, *Currier & Ives: Printmakers to the American People* (New York, 1942).
15 See Robyn Cooper, 'Millais' "The Rescue": A Painting of a Dreadful Interruption of Domestic Peace', *Art History*, IX/4 (December 1986), pp. 471–86. See also John A. Walker, *Firefighters in Art and Media: A Pictorial History* (London, 2005).
16 Manohla Dargis, 'With Many a Fear and Tear, Firefighters Prove Their Mettle', *New York Times*, 1 October 2004, p. E12.
17 William Sansom, 'The Wall', in *Fire and Water: An Anthology By Members of the NFS* [1942], ed. H. S. Ingham (Exeter, 1992), pp. 120–23.
18 *The Blitz: London's Firestorm* (TV, 2005), narrator Sophie Okonedo.
19 E. S. de Beer, ed., *The Diary of John Evelyn* (Oxford, 1955), 6 vols, III, pp. 450–51.
20 See Peter Beresford Ellis, *The Great Fire of London* (London, 1976); Stephen Porter, *The Great Fire of London* (London, 1996); Adrian Tinniswood, *By Permission of Heaven: The Story of the Great Fire of London* (London, 2003). In chapter 21, 'Painting the Town Red', in *London the Biography* (London, 2000) Peter Ackroyd gives a lively and insightful reading of the fire and its literary and artistic responses.
21 Karen Wallace, *London's Burning!* (London, 1997), p. 33.
22 Robert Latham and William Matthews, eds, *The Diary of Samuel Pepys*, VII/1666 (London, 1972), pp. 267–8.
23 Ibid., p. 268.
24 Porter, *The Great Fire of London*, p. 37.
25 Ibid., p. 39.
26 Latham and Matthews, eds, *The Diary of Samuel Pepys*, p. 271 (my emphasis).
27 Porter, *The Great Fire of London*, p. 47.
28 Latham and Matthews, eds, *The Diary of Samuel Pepys*, p. 274.
29 Ibid. See also John Adamson, 'Plucked from the Flames' (review of Tinniswood's *By Permission of Heaven*), *Sunday Telegraph*, 7 September 2003 http://global.factiva.com.ezproxy2.library.usyd.edu.au/ha/default.aspx (accessed 6 November 2005).
30 Beresford Ellis, *The Great Fire of London*; Porter, *The Great Fire of London*; Tinniswood, *By Permission of Heaven*.

31 See Tinniswood, *By Permission of Heaven.*
32 John Evelyn to Samuel Tuke, 27 September 1666, quoted ibid., p. 190.
33 Neil Hanson, *The Dreadful Judgement: The True Story of the Great Fire of London* (London and Sydney, 2001), pp. 170–71.
34 Frederick Taylor, *Dresden: Tuesday 13 February 1945* (London, 2004), p. 43. See also Michael Burleigh's review of Taylor's book in *The Guardian*, 7 February, 2004, p. 14, and David Pedlow's letter to *The Guardian*, 14 February, 2004, p. 8. See David Irving, *The Destruction of Dresden* (London, 1963). This early, now much-criticized revisionist account, grossly exaggerates the numbers of civilians killed; it cites anywhere in the region of 100,000–250,000.
35 Taylor, *Dresden*, p. xii.
36 Ibid., p. 7.
37 Ibid.
38 Gilbert, *Second World War*, p. 641.
39 Tripp quoted in Taylor, *Dresden*, p. 280.
40 Kurt Vonnegut Jr, *Slaughterhouse-Five* [1969] (London, 2000), p. 129.
41 Ibid.
42 Ibid., p. 131.
43 See Taylor, *Dresden,* chapter 7, pp. 62–75.
44 Otto Griebel, *Ich war ein Mann Strasse: Lebenserinnerungen* (Altenburg, 1995). Quoted in translation, ibid., p. 72.
45 Gottfried Semper, quoted in Wolfgang Herrmann, *Gottfried Semper: In Search of Architecture* (Cambridge, MA, 1984), p. 198. See also Gottfried Semper, *The Four Elements of Architecture and Other Writings*, trans. Harry Francis Malgrave and Wolfgang Herrmann (Cambridge, 1989).
46 'The Great Fire', *St Louis Times* (11 October 1871), p. 1.
47 Carl Smith, *Urban Disorder and the Shape of Belief: The Great Chicago Fire, the Haymarket Bomb, and the Model Town of Pullman* (Chicago and London, 1995), p. 22; David Lowe, ed., *The Great Chicago Fire In Eyewitness Accounts and 70 Contemporary Photographs and Illustrations* (New York, 1979), p. 5.
48 Smith, *Urban Disorder and the Shape of Belief*, p. 22.
49 Ibid., p. 23.
50 Elias Colbert and Everett Chamberlin, *Chicago and the Great Conflagration* (Cincinnati, OH, and New York, 1872), p. 212.
51 Ibid., pp. 216–17.
52 Ibid., p. 218.
53 *Harpers Weekly*, XV/774 (1871) p. 1004.
54 Ibid., p. 1010 (my emphasis).
55 Colbert and Chamberlin, *Chicago and the Great Conflagration*, p. 230.
56 Ibid., pp. 230–31.
57 Ibid.
58 R. Chambers, *Fires in Edinburgh* (Edinburgh, 1824), p. 54.
59 Ibid., p. 55.
60 Ibid., pp. 56–7.
61 Ibid., p. 58.
62 Ibid., p. 59.
63 See Tinniswood, *By Permission of Heaven*, p. 21.
64 Chambers, *Fires in Edinburgh*, p. 61.

65 Ibid., p. 62.
66 Quoted ibid., p. 65. See also Alexander Reid, *Aye Ready: The History of Edinburgh Fire Brigade, The Oldest Municipal Brigade in Britain* (Edinburgh, n.d.), p. 11.
67 Anon., 'Disastrous and Extensive Fire', *The Scotsman*, VIII/507 (1824), p. 1.
68 Braidwood's Report is held in the archives of the Museum of Fire, located in the Lothian and Borders Fire Brigade HQ in Edinburgh. All quotes are from this source, p. 2. I am grateful to Andrew Milne and I. McMurtie for making my research there productive and enjoyable.
69 Reid, *Aye Ready*, n.p.
70 *Weekend Australian*, 23–4 September 2006, p. 3.
71 Mark Selden, 'A Forgotten Holocaust: US Bombing Strategy, the Destruction of Japanese Cities and the American Way of War from World War II to Iraq', www.japanfocus.org/-Mark-Selden/2414 (accessed 4 April 2010).
72 Ibid.

2 Dreaming Fire / Shedding Light

1 '35,000 Revel At Festival In the Desert', *New York Times*, 6 September 2004, p. A14.
2 Alfred Birke, quoted in Frederick Taylor, *Dresden: Tuesday 13 February 1945* (London, 2004), p. 59.
3 Wally Caruana, *Aboriginal Art* [1993] (London, 2003), p. 10. See also Peter Sutton, ed., *Dreamings: The Art of Aboriginal Australia* (New York, 1988); Wally Caruana, ed., *Shadows on the Dreaming: Aboriginal Paintings in the Australian National Gallery* (Canberra, 1989); Howard Murphy, *Aboriginal Art* (London, 1998).
4 Caruana, *Aboriginal Art*, p. 10.
5 Ainslie Roberts (text) and Dale Roberts (paintings), *Shadows in the Mist* (Blackwood, South Australia, 1989), p. 18.
6 Christine Nicholls, ed., *The Magic Fire at Warlukurlangu: A Dreaming Narrative Belonging to Dolly Daniels Granites Nampijinpa* (Adelaide, 2003). All subsequent references to this Dreaming are from this source.
7 See Geoffrey Bardon (with diagrams by Judith Yan), *Papunya Tula: Art of the Western Desert* (Ringwood, Victoria, 1991); Hetti Perkins and Hannah Fink, eds, *Papunya Tula: Genesis and Genius,* exh. cat., Art Gallery of New South Wales, Sydney, 2000; Geoffrey Bardon and James Bardon, *Papunya: A Place Made after the Story: The Beginnings of the Western Desert Painting Movement* (Melbourne, 2004).
8 Bardon and Bardon, *Papunya*, p. 75.
9 Ibid., p. 60.
10 Caruana, *Aboriginal Art*, p. 117.
11 Ibid.
12 Gavin Wilson, *Fireworks: Tracing the Incendiary in Australian Art*, exh. cat., Artspace Mackay, Mackay City Council (Mackay, Queensland, 2005) p. 10.
13 Margaret Clunies-Ross, 'The Art of Fire – The Significance of Fire in the Traditional Aboriginal Performing Arts of Arnhem Land', *Fire! The Australian*

Experience: Proceedings of the 1999 Seminar, National Academies Forum: University of Adelaide, 30 September–1 October 1999 (Canberra, 2000), p. 113.
14 Ibid., p. 115.
15 'Aboriginal Fire Ceremony to Light Paralympic Flame', www.paralympic.org/paralympian/2000/04.htm (accessed 11 April 2005). In the first official event of the 2000 Olympics Arts Festival, 'The Festival of the Dreaming', aboriginal culture was also invoked. 'The Awakening Ceremony' (14 September 1997), directed by Stephen Page, was held at Bennelong Point, Sydney. It involved a 'smoking ceremony' attended by some 15,000 people. Its 'purpose was to cleanse the sight and awaken the spirits in anticipation of the festival'. Taking place at sunset, it involved singers, dancers and digeridoo players. See caption to illustration, 'The Awakening Ceremony', in Michelle Hanna, *Reconciliation in Olympism: Indigenous Culture in the Sydney Olympiad* (Sydney, 1999), p. 69.
16 See 'The Olympic Flame and Torch Relay', www.hickoksports.com/history/olflame/shtml (accessed 25 May 2005).
17 Ibid.
18 On Hestia, see Stephen M. Trazaskoma et al., eds, *Anthology of Classical Myth: Primary Sources in Translation* (Indianapolis, IN, 2004). See also Pierre Grimal, *A Concise Dictionary of Classical Mythology*, ed. Stephen Kershaw, trans. A. R. Maxwell-Hyslop [1951] (Oxford and Cambridge, MA, 1986), p. 201.
19 See M. C. Howatson, ed., 'Helios', *The Oxford Companion to Classical Literature*, 2nd edn (Oxford and New York, 1989), p. 264.
20 Gaston Bachelard, *Fragments of A Poetics of Fire*, trans. Kenneth Haltman (Dallas, TX, 1990), p. 66.
21 A 'femme fatale' in so many words, Pandora anticipates later representations of 'woman' as wily, at the most benign, and 'vagina dentata', at the most extreme. Take the Surrealists, for example: Salvador Dalí, Luis Buñuel and Paul Delvaux, among others, in whose work 'woman' is represented variously as muse, seducer and demon. See, for example, David Hopkins, *Dada and Surrealism: A Very Short Introduction* (Oxford, 2004).
22 See Arthur Cotterell and Rachel Storm, *The Ultimate Encyclopedia of Mythology* (London, 2005), p. 343.
23 On Van Baburen, see Wayne Franits, *Dutch Seventeenth-century Genre Painting: Its Stylistic and Thematic Evolution* (New Haven, CT, and London, 2004), pp. 66–8; Seymour Slive, *Dutch Painting 1600–1800* (New Haven, CT, and London, 1995), pp. 24–5.
24 See Frank Whitford, *Oskar Kokoschka: A Life* (London, 1986), pp. 192–3.
25 See Klaus Albrecht Schröder and Johann Winkler, eds, *Oskar Kokoschka*, trans. David Britt (Munich, 1991), p. 41.
26 Whitford, *Oskar Kokoschka*, p. 192.
27 See Stephen F. Eisenman, *The Abu Ghraib Effect* (London, 2007).
28 See Marjorie L. Harth, ed., *José Clemente Orozco: Prometheus* (Los Angeles and Santa Barbara, CA, 2001).
29 David W. Scorr, '*Prometheus* Revisited', ibid., pp. 27–45.
30 Orozco quoted in Harth, *José Clemente Orozco*, p. 15.
31 *The Inferno of Dante Alighieri*, trans. Ciaran Carson (London, 2002).
32 Ibid., p. 62.

33 Basil Gray, ed., *The Arts of India* (Oxford, 1981), p. 78.
34 Vidya Dehejia, *Art of the Imperial Cholas* (New York and Oxford, 1990), pp. 39–47.
35 Ananda K. Coomaraswamy, *The Dance of Siva: Essays on Indian Art and Culture* (New York, 1985), pp. 56–7.
36 Ibid., pp. 57–8.
37 Cotterell and Storm, *The Ultimate Encyclopedia of Mythology*, p. 402.
38 See Michael Kampen O'Riley, *Art Beyond the West: The Arts of Africa, India and Southwest Asia, China, Japan and Korea, The Pacific, and The Americas* (London, 2001), p. 89. See also Partha Mitter, *Indian Art* (Oxford, 2001), p. 33.
39 Nirad C. Chaudhuri, *Hinduism: A Religion to Live By* (London and Toronto, 1979), p. 72.
40 Ibid., p. 73.
41 The *Brahmana* 1.1, quoted ibid.
42 Exodus 3:2–4. On the nuanced meanings of the original Hebrew words *mikvah seneh*, 'bush that burns', see Graham Philips, *The Moses Legacy: In Search of the Origins of God* (London, Basingstoke and Oxford, 2002), pp. 277–8.
43 Cathy Frierson, *All Russia is Burning: A Cultural History of Fire and Arson in Late Imperial Russia* (Seattle, WA, and London, 2002) p. 25.
44 Quoted ibid., pp. 25–6.
45 Daniel 3:21.
46 Colour print (1775), illustrated in Martin Butlin, *William Blake*, exh. cat., Tate Gallery, London (1978), cat. no. 91, p. 61.
47 T.S.R. Boase, *Nebuchadnezzar* (London, 1972).
48 See Barry Pearce et al., *Arthur Boyd: Retrospective*, exh. cat., Art Gallery of New South Wales, National Gallery of Victoria, Tasmanian Museum and Art Gallery and Art Gallery of Western Australia (Sydney, 1994), pp. 25–6.
49 This work is illustrated in *Fine Australian Paintings*, sale cat., sale by Sotheby's Australia, Sydney (22 November 1992), item no. 286 (colour); see also *Australian Modernism: The Complexity and the Diversity*, exh. cat., Lauraine Diggins Gallery, Melbourne (1992), p. 26 (colour). For these references, I am grateful to Steven Miller, Archivist and Acting Head Librarian, Art Gallery of New South Wales, Sydney.
50 Daniel 3:21.
51 Boase, *Nebuchadnezzar*, p. 20.
52 See, for example, Ursula Hoff, *The Art of Arthur Boyd* (London, 1986), p. 58; Darleen Bungey, *Arthur Boyd: A Life* (Crows Nest, New South Wales, 2007), p. 435.
53 On Macleod's fire figure series, see Craig Judd and Amanda Lawson, 'Introduction', in *Euan Macleod, Napoleon Reef: Paintings 1992–2002*, exh. cat., Bathurst Regional Art Gallery, Bathurst (2002), p. 6.
54 Daniel 4:10–14.
55 Hoff, *The Art of Arthur Boyd*, p. 59.
56 Pearce, *Arthur Boyd: Retrospective*, p. 26.
57 Born in Cappagh County, Ireland, in 1916, Tyrrell had suffered abuse at St Joseph's Industrial School in Letterfrack run by the Christian Brothers. Becoming a manic-depressive, he would later enlist in the British Army, ending up as a prisoner of war at Stalag XIB. Returning to London in 1945, his last

twenty or so years were marked by his experience of virulent anti-Irish sentiment. All this and more has been brought together in a recently published memoir based on a rediscovered manuscript by Tyrrell: Peter Tyrrell, *Founded on Fear* (Dublin, 2006). See also John Healy, 'From Beyond the Grave', *The Mayo News* (15 November 2006), www.mayonews.ie/index.php?Itemid=35&id=620 (accessed 7 August 2009).

58 See 'Self-Immolation in Vietnam', www.francesfarmersrevenge.com/stuff/archive/oldnews2/sel (accessed 8 February 2007).

59 Haberstam, quoted in Russell T. McCutcheon, *Manufacturing Religion: The Discourse on Sui Generis Religion and the Politics of Nostalgia* (Oxford and New York, 1997), p. 168. I am indebted to the section 'Representing Vietnamese "Self-Immolations"', pp. 167–77.

60 Carl-Martin Edsman, from an article on fire in Mircea Eliade, ed., *Encyclopedia of Religion*, vol. v (New York, 1987), quoted in McCutcheon, *Manufacturing Religion*, p. 171.

61 Thich Nhat Hanh, *Vietnam: Lotus in a Sea of Fire* (New York, 1967), p. 16, quoted in McCutcheon, *Manufacturing Religion*, p. 174.

62 See Xavier Costa, 'Festive Identity: Personal and Collective Identity in the Fire Carnival of the "Fallas" (València, Spain)', *Social Identities*, viii/2 (2002), pp. 321–45; Simon Wallis, 'Fire Fiesta', http://homepage.mac.com/simon.wallis/site/newspapers/fallas.htm (accessed 8 January 2006); Tom Little, 'Las Fallas Festival', *Out of Town*, www.barcelonametropolitan.com/Article.aspx?TabID=2&MenuID=3& (accessed 8 January 2006).

63 Quoted in Costa, 'Festive Identity', p. 340.

64 Wallis, 'Fire Fiesta'.

65 'Naga Fireballs Festival 28 October 2005', www.discoverythailand.com/nage_fireballs_festival_2005.asp (accessed 23 December 2005). See also Jason Gagliardi, 'Behind the Secret of the Naga's Fire', www.time.com/magazine/article/0,9171.391567,00.html (accessed 23 December 2006).

66 Gary Walsh, 'River of Fire', *Sydney Morning Herald*, 13–14 March 2004, p. 5. See also Gary Walsh, 'Great Balls of Fire', *Sydney Morning Herald*, www.smh.com.au/articles/2003/11/14/1068674369352.html (accessed 14 January 2006).

67 Walsh, 'River of Fire'.

68 Jira Malikul, quoted in Pattana Kitiarsa, 'Faith and Films: Countering the Crisis of Thai Buddhism from Below', *Asian Journal of Social Sciences*, xxxiv/2 (2009), p. 279.

69 Ibid., pp. 278–82.

70 Phrakhu Pichai Kitjaton, quoted in Gagliardi, 'Behind the Secret of the Naga's Fire'.

71 Anthony-Haden Guest, 'In Defense of Burning Man', *Art in America* (June/July 2005), p. 114. See also Brian Doherty, *This is Burning Man: The Rise of a New American Underground* (New York, 2004). For an unfavourable review of this book, see Dave Itzkoff, 'The Rite of Summer', *New York Times Book Review* (19 September 2004), p. 14.

72 See 'Festival of Fire Draws Record Crowds', *Los Angeles Times*, 6 September 2004, p. A22.

73 See http://afterburn.burningman.com/08/ (accessed 14 May 2008).

74 See Allan Brown, *Inside the 'Wicker Man': The Morbid Ingenuities* (London, 2000).
75 Harvey quoted in Guest, 'In Defense of Burning Man'.
76 '35,000 Revel At Festival In the Desert'.
77 Ibid.
78 The photograph of Jimi Hendrix on stage setting fire to his guitar was part of a large private collection of Hendrix memorabilia auctioned in 2004; it is reproduced in *The Times*, 11 August 2004, p. 7.

3 The Intimate Flame

1 'Fire Talk', www.i2k.com/~suzanne/page8.htm (accessed 27 November 2002).
2 Ibid. In correspondence with the author (18 January 2006), Adams has said: 'Fire is one of my favourite elements – I have to have a real fireplace in my house at least. I moved away from the home in the woods with the fire pit from where I painted *Fire Talk*. I burn wood in my fireplace now and watch its ever-changing artwork.'
3 *Sydney Morning Herald*, 2 January 2006, p. 5.
4 Luis Fernández-Galiano, *Fire and Memory: On Architecture and Energy*, trans. Gina Carino (Cambridge, MA, 2000), p. 6; paraphrasing Reyner Banham, *The Architecture of the Well-Tempered Environment* (London, 1969), p. 19.
5 Fernández-Galiano, *Fire and Memory*, p. 7 (my emphasis).
6 Vitruvius, *Ten Books on Architecture*, trans. Ingrid D. Rowland (Cambridge, 1999), p. 34.
7 Ibid., p. 35.
8 Leon Battista Alberti, quoted in Fernández-Galiano, *Fire and Memory*, pp. 9–10.
9 Leon Battista Alberti, *On the Art of Building in Ten Books*, trans. Joseph Rykwert, Neil Leach and Robert Tavernor (Cambridge, MA, 1988), p. 355.
10 Gaston Bachelard, *The Psychoanalysis of Fire* [1938], trans. Alan C. M. Ross (Boston, MA, 1968), pp. 14–15.
11 Georges Duby, ed., *A History of Private Life, II: Revelations of the Medieval World*, trans. Arthur Goldhammer (Cambridge, MA, 1987), p. 426.
12 Fernand Boudel, *The Stuctures of Everyday Life*, trans. Siân Reynolds, vol. I (London, 1981), p. 298.
13 Duby, *A History of Private Life*, p. 194.
14 See www.getty.edu/art/gettyguide/artObjectDeaials?arto (accessed 9 February 2004).
15 See Jeroen Giltaij et al., *Senses and Sins: Dutch Painters of Daily Life in the Seventeenth Century*, exh. cat., Museum Boijmans Van Beuningen, Rotterdam, and Städelsches Kunstinstituit und Städtische Galerie, Frankfurt am Main (2005), p. 66.
16 Friso Lammertse, ibid., p. 157.
17 Ibid., Lammertse, citing Cesare Ripa, *Iconologia, Of uytbeeldingen des Verstands*, trans. D. P. Pers (Amsterdam, 1744), p. 358.
18 See 'Review', in *National Arts Collections Fund*, National Gallery of Scotland, Edinburgh (1990) p. 195.

19 Hugh Brigstocke, *Italian and Spanish Paintings in the National Gallery of Scotland* (Edinburgh, 1993), pp. 77–80.
20 Peter Schjeldahl, 'A Woman's Work', *The New Yorker* (29 June 2009), p. 81.
21 Ann Sutherland Harris, *Seventeenth-Century Art and Architecture* (London, 2005), p. 359.
22 See the excellent John Emsley, *The Shocking History of Phosphorus: A Biography of the Devil's Element* (London, 2000). See also www.rpsgb.org.uk/members/museum/musexhibwajo.htm (accessed 30 March 2005); and 'The History of Matches', http://inventors.about.com/library/inventors/blmatch.htm (accessed 30 March 2005).
23 See 'John French – The Art of Distillation', book 5 of 6 (London, 1651), www.levity.com/alchemy/jfren_ar.html (accessed 20 July 2006).
24 Ibid.
25 See Science Museum, London, www.ingenious.org.uk/See/?target=SeeLarge&ObjectID={7 (accessed 27 April 2006).
26 Ibid.
27 Science Museum, London, www.ingenious.org.uk/See/?target=SeeLarge&ObjectID={6... (accessed 27 April 2006).
28 See 'The History of Matches'.
29 At http://librarythinkquest.org/C0122240/solidrockhistory.htm (accessed 20 April 2006).
30 Science Museum, London, www.ingenious.org.uk/See/?target=SeeLarge&ObjectID={...
31 *Oxford Dictionary of English*, 2nd edn (Oxford, 2003), p. 1042.
32 Cited in Peter E. Childs, 'Phosphorus: From Urine to Fire', www.ul.ie/~childsp/CinA/Issue61/ToC25_Phosphorous.html (accessed 27 April 2006).
33 See www.rpsgb.org.uk/members/museum/musexhibwajo.htm (accessed 30 March 2005).
34 I thank the Australian writer Evelyn Juers, who drew my attention to Hoffmann's tale. Juers makes some keen observations on fire and children's literature in the introduction to her 'Janis Joplin and the Sisters of Excess', *Heat 7* (Sydney, 1998).
35 Heinrich Hoffmann, 'The Dreadful Story of Harriet and the Matches', *Struwwelpeter: Merry Stories and Funny Pictures* [German, 1844; English, 1848] (New York, n.d.), www.gutenberg.org/files/12116/12116-h/12116-h.htm (accessed 30 April 2005).
36 *The Times*, 8 June 1867, p. 10. See also anon., *Bryant & May: Making Matches, 1861–1961* (London [1961]), n.p.
37 Ibid.
38 Ibid.
39 All quotes are taken from Hans Christian Andersen, 'The Little Match-Seller' (1846), www.hca.gilead.org.il/li_match.html (accessed 23 June 2006).
40 At www.rhymes.org.uk/jack_be_nimble.htm (accessed 19 May 2006).
41 See Rob Ossian, 'Pirates, Privateers, Corsairs, Explorers and More', www.geocities.com/Athens/7012/bios/index.htn (accessed 29 May 2006). A specialist on piracy, Ossian has written to the author (1 June 2006), saying: 'There are many uses of the term "Black Jack" . . . but no historically influential pirates to my knowledge have ever worn that moniker. I would personally

lean to the likelihood that the origin was . . . in reference to a leather-covered bludgeon with a short, flexible shaft or strap, used as a hand weapon since coming up with a Black Jack . . . clobbers your opponent. These type of bludgeing [sic] tools have been around since before the European renaissance, and therefore certainly precede the game.

Further, when the English would troll the wharf-side pubs for drunks to impress into service on their naval vessels they would commonly use "black jacks" to knock them unconscious, making them a vile and hated tool to any sailor and doubtful one that they would want as a nickname.'

42 Chris Roberts, *Heavy Words Lightly Thrown: The Reason Behind the Rhyme* (London, 2003), p. 8.

43 *The Illustrated Gaelic-English Dictionary* [1901–11], ed. Edward Dwelly (Glasgow, 1973) pp. 81–2.

44 See 'High Jump', www.rooneydesign.com/JackBeNimble.html (accessed 19 May 2006).

45 Thomas Wright, *The Romance of the Lace Pillow* (Carlton, 1982), p. 196.

46 Ibid. According to Alexandra MacCulloch, Keeper of Art, Clothing and Textiles, Buckinghamshire County Museum: 'Suffolk does not have the same tradition of lace making as other parts of the UK like Bucks and Honiton and the majority of children learning to make lace would have been girls . . . I am not sure how widely believed this derivation is for the nursery rhyme.' (Correspondence with the author, 4 July 2006.)

47 Anne Buck, *Thomas Lester, his Lace and the East Midlands Industry 1820–1905*, (Bedford, 1981), p. 4.

48 C. Channer and M. E. Roberts, *Lace-Making in the Midlands* (London, 1900), pp. 31–7.

49 Buck, *Thomas Lester*, p. 5. See also, 'A Calender of Lacemakers Celebrations', http://bobbins.lacefairy.com/BobbinMuseum/calender.html (accessed 22 May 2006).

50 Vicki Elliot, 'Nowruz eid-i shoma mobaraki! For Iranians, fire jumping heralds a fresh start on the new year', www.sfgate.com/cgibin/article.cgi?f=c/a/200403/13/Ho (accessed 29 May 2006).

51 Paliman Salimpour, quoted ibid.

52 Iona and Peter Opie, *The Oxford Dictionary of Nursery Rhymes* (Oxford, 1951), p. 268.

53 Bessie Ling, interviewed by Ginette Dunn (2 October 1974), in Dunn, *The Fellowship of Song: Popular Singing Traditions in East Suffolk* (London, 1980), p. 115.

54 Ben Ling, interviewed by Dunn (2 July 1975), in Dunn, *The Fellowship of Song: Popular Singing Traditions in East Suffolk*, p. 118.

55 Bachelard, *The Psychoanalysis of Fire*, p. 43.

56 See the brochure accompanying the CD, *Jane Eyre* (Chandos Records Ltd, England, 2002).

57 All references to the novel are from Dr Sally Minogue, 'Introduction' and 'Notes' in Charlotte Brontë, *Jane Eyre* (Ware, 1999). See Patricia Ingram, *The Brontës* (Oxford, 2006). See also Heather Glen, ed., *The Cambridge Companion to the Brontës* (Cambridge, 2002).

58 See Brian Scherle, 'The Heart is the Object of Fire: Fire and Passion in *Jane*

Eyre', www.umd.umich.edu/casl/hum/eng/calsses/434/carweb/FIRE.htm (accessed 20 April 2006).
59 Brontë, *Jane Eyre*, p. 30.
60 Ibid., p. 99.
61 Ibid., p. 102.
62 Ibid.
63 Ibid.
64 Ibid., p. 129.
65 Ibid., p. 130.
66 Ibid.
67 For a short but teasing cultural history of the bed, see Anthony Burgess, *On Going to Bed* (London, 1982).
68 Minogue, 'Introduction', p. xvi.
69 Brontë, *Jane Eyre*, p. 334.
70 Ibid., p. 339.
71 *Robert Frost: Collected Poems, Prose, and Plays*, ed. Richard Poirier and Mark Richardson (New York, 1995), p. 204.
72 Jay Parini, *Robert Frost: A Life* (London, 2001), p. 198.
73 Ibid.
74 Brontë, *Jane Eyre*, p. 371.
75 Ibid., p. 376.
76 Ibid., p. 377.

4 Playing with Fire

1 Ahmed Nassir Juma Bhalo, 'The Fire', in *Nine Faces of Kenya: An Anthology*, ed. Elspeth Huxley (London, 1991), p. 404.
2 Neil Watson, *Burn: Artists Play with Fire*, exh. cat., Norton Museum of Art and Columbia Museum of Art (West Palm Beach, FL, 2001), p. 11.
3 The critic Jerry Saltzer has said that 'For Fischli and Weiss play is the primary process involved in creativity.': Saltzer, 'Notes on a Film', *Arts Magazine* (April 1988), pp. 11–12; Arthur C. Danto has written profitably on the idea of play's transformative character in their work: see Robert Fleck, Beate Söntgen and Arthur C. Danto, *Peter Fischli and David Weiss: In a Restless World*, exh. cat., Walker Art Center, Minneapolis (1966), pp. 95–113.
4 Peter Fischli and David Weiss, *The Way Things Go* (1987) 16mm film, 30 mins duration. AT&C Film AG Zurich Production in association with Alfred Richterich. References in my text are to the DVD, *The Way Things Go* (2002).
5 See Robert Fleck, Beate Söngten and Arthur C. Danto, *Peter Fischli, David Weiss* (London and New York, 2005), pp. 46–9.
6 At www.buchiglas.ch/english/buchiglas/cgm (accessed 23 July 2007).
7 Dr Pietro Maggi, Uster City Archivist, correspondence with the author (2 August 2007).
8 See Danto, *Peter Fischli and David Weiss: In a Restless World*, p. 101, n. 7.
9 Virginia Rutledge, 'Suddenly It All Makes Sense', *Art in America* (June 1997), p. 105.
10 Joanna Ball, 'The Ritual of the Necklace'. Research report written for the

Centre for the Study of Violence and Reconciliation (South Africa), March 1994, www.csvr.org.za/wits/papers/papball.htm (accessed 14 April 2010).

11 Ibid., p. 3.

12 Arthur C. Danto writes briefly about the parallels between the two works in *Peter Fischli and David Weiss*, pp. 100–101.

13 See Jean Tinguely and Pontus Hultén, *Jean Tinguely: Meta* (London and Boston, 1976); Pontus Hultén, *Jean Tinguely: A Magic Stronger than Death* (Boston, 1987).

14 Calvin Tomkins, 'Jean Tinguely – Beyond the Machine', *The New Yorker* (10 February 1962): extracts reproduced in Barry Schwabsky and Calvin Tomkins, *Michael Landy H2YN* (New York, 2007), pp. 85–93, a book published to coincide with the exhibition of the same name at Alexander and Bonin, New York. All references to Tomkins's essay are from this source.

15 Tinguely, quoted ibid., p. 93.

16 See *Michael Landy H2YN*.

17 Michael Landy, *Man in Oxford Street is Auto-Destructive*, exh. cat., Sherman Galleries, Sydney (2007).

18 Peter Hill, quoted ibid., n.p.

19 Martin Honert, quoted by Boris Groys in *Martin Honert*, exh. cat., Matthew Marks Gallery, New York (2002), n.p.

20 See the full-page colour illustration, ibid.

21 Mark Pimlott, 'Human World', *Frieze* (March–April 1994), p. 58.

22 Boris Groys, 'Mind's Eye Views: Boris Groys Talks with Martin Honert', *Artforum*, XXXIII/6 (February 1995), p. 59.

23 Ruscha, quoted in Lynne Zelevansky, 'An Interview with Ed Ruscha', in *Magritte and Contemporary Art: The Treachery of Images*, ed. Stephanie Barron and Michel Draguet with the assistance of Sara Cochran, exh. cat., Los Angeles County Museum of Art, Los Angeles (2007), p. 137.

24 Kerry Brougher et al., *Ed Ruscha* (Washington, DC, and Oxford, 2000), p. 161.

25 Ibid., p. 151.

26 Mary Richards, *Ed Ruscha* (London, 2008), pp. 60–61.

27 Ruscha, quoted ibid., p. 61.

28 See 'The Lifeline', the title of a lecture Magritte gave on 20 November 1938 at the Musée Royal des Beaux-Arts, Antwerp; reproduced in Harry Torczyner, *Magritte: Ideas and Images*, trans. Richard Miller (New York, 1979), pp. 118–25.

29 From a letter Magritte sent to André Breton, cited in David Sylvester, ed., *René Magritte: Catalogue Raisonné*, vol. III (London, 1993), p. 190.

30 Ibid., fig. a.

31 Sylvester, *Catalogue Raisonné*, I, no. 234, p. 284.

32 Ibid., no. 235.

33 Sylvester, *Catalogue Raisonné*, II, no. 390, pp. 212–13.

34 Ibid., no. 698, p. 441.

35 See Gilbert Perlein and Bruno Corra, *Yves Klein: Long Live the Immaterial*, exh. cat., Musée d'Art Moderne et d'Art Contemporain, Nice, and Museo Pecci, Prato (New York, 2001).

36 Reproduced in Peter Noever and François Perrin, eds, *Yves Klein: Air Architecture* (Los Angeles, 2004), pp. 35–45. This book was published on the occasion of the exhibition of the same name at the MAK Center for Art

and Architecture, Los Angeles (13 May – 29 August 2004).
37 Hannah Weitemeier, *Yves Klein, 1928–1962: International Klein Blue* (Cologne, 1995), p. 76.
38 Gaston Bachelard, *The Psychoanalysis of Fire*, trans. Alan C. M. Ross (Boston, 1964), p. 7.
39 See Pierre Restany, *Yves Klein: Fire at the Heart of the Void* (Putnam, CT, 2005), pp. 2–3.
40 Noever and Perrin, *Yves Klein: Air Architecture*, p. 11.
41 Brian O'Doherty, *Inside the White Cube: The Ideology of the Gallery Space* (London, 1976).
42 Sylvère Lotringer, 'Travelling Through the Void', in *Yves Klein: Air Architecture*, p. 126.
43 See Sidra Stich, *Yves Klein* (Ostfildern-Ruit, 1994), p. 120.
44 Restany, *Yves Klein*, p. 3.
45 This matter is discussed by Yves-Alain Bois, 'Klein's Relevance for Today', *October* 119 (Winter 2007), p. 89.
46 Restany, *Yves Klein*, p. 3.
47 Ibid., p. 7.
48 Noever and Perrin, *Yves Klein: Air Architecture*, p. 106.
49 Ibid., p. 44 (my emphasis).
50 Ibid., p. 41.
51 Restany quoted in Weitemeier, *Yves Klein 1928–1962*, p. 71.
52 Stich, *Yves Klein*, p. 223.
53 Weitemeier, *Yves Klein 1928–1962*, p. 76.
54 Klein quoted in Anja Zimmermann, '"Sorry for Having to Make You Suffer": Body, Spectator, and the Gaze in the Performances of Yves Klein, Gina Pane, and Orlan', *Discourse*, XXIV/3 (2002), pp. 27–46, http://muse.jhu.edu/journals/discourse/v024/24.3zimmermann.html (accessed 22 March 2009).
55 Ibid.
56 See Stich, *Yves Klein*, p. 226.
57 Bois, 'Klein's Relevance for Today', p. 93.
58 Quoted by Guy Brett in Olga M. Viso, *Ana Mendieta: Earth Body: Sculpture and Performance, 1972–1985* (Ostfildern-Ruit, 2004), p. 186. This book was published in conjunction with the exhibition of the same name organized by the Hirshhorn Museum and Sculpture Garden, Smithsonian Institution, Washington: Whitney Museum of American Art, New York (1 July – 1 September 2004); Hirshhorn Museum and Sculpture Garden, Washington (14 October 2004 – 2 January 2005); Des Moines Art Center (25 February – 22 May 2005); Miami Art Museum (7 October 2005 – 15 January 2006).
59 See Chrissie Iles, in Viso, *Ana Mendieta: Earth Body*, p. 210.
60 Olga Viso, *Unseen Mendieta: The Unpublished Works of Ana Mendieta* (Munich, Berlin, London and New York, 2008), p. 153.
61 Guy Brett, in *Ana Mendieta: Earth Body*, p. 186.
62 Viso, *Unseen Mendieta*, various pages.
63 Ibid., pp. 186–9.
64 Ibid., p. 196.
65 Ibid., p. 162.
66 Ibid., p. 155.

67 See Caroline Collier and Stephen Foster, 'Foreword', in *Gina Pane* (Southampton, 2002), p. 7. This publication was produced on the occasion of the exhibition of the same name at John Hansard Gallery, University of Southampton (27 November 2001 – 19 January 2002); Arnolfini, Bristol (23 February – 14 April 2002); and a display at Tate Modern, London (April–October 2002).
68 Ibid., p. 19.
69 Cynthia Carr, *On Edge: Performance at the End of the Twentieth Century* (Hanover and London, 1993), p. 16.
70 Julian Hountou, 'Le corps au mur: La méthode photographique de Gina Pane', *Etudes Photographiques,* no. 8 (November 2000), http://etudesphotographiques.revues.org/index229 (accessed 20 February 2010).
71 Pane, cited by Anja Zimmermann, '"Sorry for Having to Make You Suffer"' (accessed 29 April 2010).
72 See Roselee Goldberg, *Performa: New Visual Art Performance* (New York, 2005), pp. 117–21.
73 Pane, quoted by Kathy O'Dell, *Contract with the Skin: Masochism, Performance Art, and the 1970s* (Minneapolis, MN, and London, 1998), p. 45.
74 Mike Parr, 'Some Notes on my Work 1970–1974', 1974, revised with an addendum by the artist 16 November 2007, in *Mike Parr Performances 1971–2008* (Melbourne, 2008), n.p.
75 Terry Smith, 'Live Art's Effects and Defects', *The Review* (17–23 June 1972), p. 996, quoted by David Bromfield, *Identities: A Critical Study of the Work of Mike Parr 1970–1990* (Nedlands, Western Australia, 1991), p. 38.
76 Ibid., p. 38.
77 *Rules & Displacement Activities: Part 1, Mike Parr Performances,* p. xiv, illus. 59.
78 Illustrated ibid., p. xvii.
79 Bunkier Sztuki, Kraków, recently held a retrospective of Bereś: 'Jerzy Bereś: Sztuka zgina zycie / Art bends Life' (12 June – 9 September 2007). It was curated by Maria Anna Potocka. The exhibition catalogue has the same name and was published by Galeria Sztuki Wspolczesnej Bunkier Sztuki, Kraków (2007).
80 Piotr Piotrowski, *In the Shadow of Yalta: Art and the Avant-Garde in Eastern Europe, 1945–1989,* trans. Anna Brzyski (London, 2009), p. 380.
81 Ibid., pp. 381–2.
82 Ibid., p. 389.
83 Correspondence with the author (24 May 2010).
84 Klara Kemp-Welch, 'Jerzy Bereś's Radical Expansion of the Readymade Event', unpublished draft, pp. 4–5. I am indebted to Kemp-Welch for providing me with this draft of her article, which went on to be published in the catalogue accompanying the Bunkier Sztuki exhibition of Bereś. The essay was an extended version of a paper Kemp-Welch had given at the Social East Forum seminar no. 3 on 'Art and Revolution', at Manchester Art Gallery (3 February 2007). That paper was titled 'Readymade Rituals: Jerzy Bereś in Dialogue with Marcel Duchamp'.
85 Ibid., p. 5.
86 'Jan Dibbets: TV as a fireplace', www.medienkunstnetz.de/works/tv-as-a-fireplace/ (accessed 20 January 2010).
87 Jan Dibbets interviewed by Catherine Mosley in *Conception: Conceptual Documents 1968 to 1972,* exh. cat., Norwich Gallery, Norwich School of Art

and Design (Norwich, 2001), p. 125. My thanks to Kevin Stahl, Archivist at Gladstone Gallery, New York, for drawing my attention to this reference. See also Lucy Lippard, *Six Years: The Dematerialization of the Art Object* (New York, 1973), p. 133.

88 From an interview with Catherine Kinley, in Barbara Einzig, ed., *Thinking About Art: Conversations with Susan Hiller* (Manchester and New York, 1996), p. 89. See also James Lingwood, ed., *Susan Hiller: Recall: Selected Works 1969–2004* (Gateshead, 2004). This book was published on the occasion of the exhibition of the same name at Baltic, Gateshead (1 May – 18 July 2004); Museu de Arte Contemporânea de Serralves, Porto (15 October 2004 – 9 January 2005); Kunsthalle Basel (30 January – 27 March 2005).

89 Lucy Lippard, 'Preface', in *Thinking About Art*, p. 91.

Postscript: '*A Dreadful Prankster*'

The subtitle of this postscript comes from curator Neil Watson's memorable description of fire, from *Burn: Artists Play With Fire*, exh. cat., Norton Museum of Art, West Palm Beach, FL (2001), p. 11.

1 See Mphuthumi Ntabeni, 'Burning Issues', *G21 Africa #438, Tenth Anniversary Special Edition 1996–2006*, http://g21.net/africa118.html (accessed 14 April 2010).

2 Bryan Walsh, 'Why Global Warning May Be Fueling Australia's Fires', *Time* (9 February 2009), http:/www.time.com/time/health/article// 0,8599,1878220,00.html (accessed 10 May 2010).

3 Ibid.

4 Nicky Phillips, 'Singed but Safe: Test House Survives the Ravages of "Extreme"' Bushfire', *Sydney Morning Herald*, 17–18 April 2010, p. 13.

5 Mark Denning, dir., DVD *Fireplace* (2001). See www.dvdproductions.com.au.

BIBLIOGRAPHY

'35,000 Revel At Festival In the Desert', *New York Times*, 6 September 2004, p. A14

'Aboriginal Fire Ceremony to Light Paralympic Flame', www.paralympic.org/paralympian/2000/04.htm

Ackroyd, Peter, *London the Biography* (London, 2000)

Adams, Suzanne, 'Fire Talk', www.i2k.com/~suzanne/page8.htm

Adamson, John, 'Plucked from the Flames', review of Adrian Tinniswood, *By Permission of Heaven*, *Sunday Telegraph*, 7 September 2003, document STEL000020030907dz970003p, global.factiva.comezproxy2.library.usyd.edu.au/ha/default.aspx

Alberti, Leon Battista, *On the Art of Building in Ten Books*, trans. Joseph Rykwert, Neil Leach and Robert Tavernor (Cambridge, MA, 1988)

Alighieri, Dante, *The Inferno of Dante Alighieri*, trans. Ciaran Carson (London, 2002)

Australian Modernism: The Complexity and the Diversity, exh. cat., Lauraine Diggins Gallery, Melbourne (1992)

Bachelard, Gaston, *Fragments of a Poetics of Fire*, trans. Kenneth Haltman (Dallas, TX, 1990)

——, *The Psychoanalysis of Fire* [1938], trans. Alan C. M. Ross (Boston, MA, 1968)

Ball, Joanna, 'The Ritual of the Necklace'. Research report written for the Centre for the Study of Violence and Reconciliation [South Africa] March 1994, www.csvr.org.za/wits/papers/papball.htm

Banham, Reyner, *The Architecture of the Well-Tempered Environment* (London, 1969)

Bardon, Geoffrey (with diagrams by Judith Yan), *Papunya Tula: Art of the Western Desert* (Ringwood, Victoria, 1991)

——, and James Bardon, *Papunya: A Place Made after the Story: The Beginnings of the Western Desert Painting Movement* (Melbourne, 2004)

Benezra, Neal, and Kerry Brougher with a contribution by Phyllis Rosenzwieg, *Ed Ruscha* (Washington, DC, and Oxford, 2000)

Bhalo, Ahmed Nassir Juma, 'The Fire', in *Nine Faces of Kenya: An Anthology*, ed. Elspeth Huxley (London, 1991)

Boase, T.S.R., *Nebuchadnezzar* (London, 1972)

Boehm, Lisa Krissoff, *Popular Culture and the Enduring Myth of Chicago, 1871–1968* (London, 2004)

Bois, Yves-Alain, 'Klein's Relevance for Today', *October*, 119 (Winter 2007), pp. 75–93

Braudel, Fernand, *The Structures of Everyday Life*, trans. Siân Reynolds, vol. I (London, 1981)

Brigstocke, Hugh, *Italian and Spanish Paintings in the National Gallery of Scotland* (Edinburgh, 1993)

Bromfield, David, *Identities: A Critical Study of the Work of Mike Parr 1970–1990* (Nedlands, Western Australia, 1991)
Brown, Allan, *Inside the 'Wicker Man': The Morbid Ingenuities* (London 2000)
Bryant & May: Making Matches, 1861–1961 (London [1961])
Buck, Anne, *Thomas Lester, his Lace and the East Midlands Industry 1820–1905* (Bedford, 1981)
Buckman, David, 'Obituary: Paul Dessau', *The Independent*, 11 November 1999, p. 7
Bungey, Darleen, *Arthur Boyd: A Life* (Crows Nest, New South Wales, 2007)
Burgess, Anthony, *On Going to Bed* (London, 1982)
Burleigh, Michael, review of Frederick Taylor, *Dresden: Tuesday 13 February 1945*, *The Guardian*, 7 February 2004
'Bushfires in Victoria', http://guides.slv.vic.gov.au/content.php?pid=36394&sid=267991
Butlin, Martin, *William Blake*, exh. cat., Tate Gallery, London (1978)
'A Calender of Lacemakers Celebrations', http://bobbins.lacefairy.com/BobbinMuseum/calender.html
'California Pins Hope on Damp Ocean Breezes', *The Eagle Tribune* (28 October 2003), www.ecnnews.com/cgi-bin/05/etarchivestory.pl?200310+fin
Carr, Cynthia, *On Edge: Performance at the End of the Twentieth Century* (Hanover and London, 1993)
Caruana, Wally, *Aboriginal Art* [1993] (London, 2003)
——, ed., *Shadows on the Dreaming: Aboriginal Paintings in the Australian National Gallery* (Canberra, 1989)
Chambers, R., *Fires in Edinburgh* (Edinburgh, 1824)
Channer, C. C., and M. E. Roberts, *Lace-Making in the Midlands* (London, 1900)
Chapin, John R., 'Letter', *Harper's Weekly*, XV/775 (1871), p. 1010
Chaudhuri, Nirad C., *Hinduism: A Religion to Live By* (London and Toronto, 1979)
Childs, Peter E., 'Phosphorus: From Urine to Fire', www.ul.ie/~childsp/CinA/Issue61/ToC25_Phosphorous.html
Clunies-Ross, Margaret, 'The Art of Fire – The Significance of Fire in the Traditional Aboriginal Performing Arts of Arnhem Land', *Fire! The Australian Experience: Proceedings of the 1999 Seminar, National Academies Forum: The University of Adelaide, 30 September – 1 October 1999* (Canberra, 2000)
Colbert, Elias, and Everett Chamberlin, *Chicago and the Great Conflagration* (Cincinnati, OH, and New York, 1872)
Collier, Caroline, and Stephen Foster, *Gina Pane* (Southampton, 2002)
Conception: Conceptual Documents 1968 to 1972, exh. cat., Norwich Gallery, Norwich School of Art and Design (Norwich, 2001)
Cooper, Robyn, 'Millais' "The Rescue": A Painting of a Dreadful Interruption of Domestic Peace', *Art History*, IX/4 (December 1986), pp. 471–86
Coomaraswamy, Ananda K., *The Dance of Siva: Essays on Indian Art and Culture* (New York, 1985)
Costa, Xavier, 'Festive Identity: Personal and Collective Identity in the Fire Carnival of the "Fallas" (València, Spain)', *Social Identities*, VIII/2 (2002), pp. 321–45
Cotterell, Arthur, and Rachel Storm, *The Ultimate Encyclopedia of Mythology* (London, 2005)

Dargis, Manohla, 'With Many a Fear and Tear, Firefighters Prove Their Mettle', *New York Times*, 1 October 2004, p. E12.
De Bolla, Peter, *The Discourse of the Sublime: Readings in History, Aesthetics and the Subject* (London, 1989)
Dehejia, Vidya, *Art of the Imperial Cholas* (New York and Oxford, 1990)
Denning, Mark, dir., DVD *Fireplace* (2001), www.dvdproductions.com.au
Dibbets, Jan, 'TV as a Fireplace', www.medienkunstnetz.de/works/tv-as-a-fireplace/
'Disastrous and Extensive Fire', *The Scotsman*, VII/507 (1824), p. 1
Doherty, Brian, *This is Burning Man: The Rise of a New American Underground* (New York, 2004)
Duby, Georges, ed., *A History of Private Life, II: Revelations of the Medieval World*, trans. Arthur Goldhammer (Cambridge, MA, 1987)
Dunn, Ginette, *The Fellowship of Song: Popular Singing Traditions in East Suffolk* (London, 1980)
Einzig, Barbara, ed., *Thinking About Art: Conversations with Susan Hiller* (Manchester and New York, 1996)
Eisenman, Stephen F., *The Abu Ghraib Effect* (London, 2007)
Elliot, Vicki, '*Nowruz eid-i shoma mobaraki!* For Iranians, Fire Jumping Heralds a Fresh Start on the New Year', www.sfgate.com/cgi-bin/article.cgi?f=c/a/200403/13/Ho…
Ellis, Peter Beresford, *The Great Fire of London* (London, 1976)
Emsley, John, *The Shocking History of Phosphorus: A Biography of the Devil's Element* (London, 2000)
Evelyn, John, *The Diary of John Evelyn, III: 1650–1672*, ed. E. S. de Beer (Oxford, 1955)
'The Expulsion from Paradise', www.getty.edu/art/collections/objects/o2892.html
Fernández-Galiano, Luis, *Fire and Memory: On Architecture and Energy*, trans. Gina Carino (Cambridge, MA, 2000)
'Festival of Fire Draws Record Crowds', *Los Angeles Times*, 6 September 2004, p. A22
Fischli, Peter, and David Weiss, *The Way Things Go* [1987], A T & C Film AG Zurich Production in association with Alfred Richterich, Icarus Films (1990)
Fleck, Robert, Beate Söngten and Arthur C. Danto, *Peter Fischli David Weiss* (London and New York, 2005)
——, *Peter Fischli and David Weiss: In a Restless World*, exh. cat., Walker Art Center, Minneapolis (1966)
Foucault, Michel, *This is Not a Pipe*, trans. James Harkness (Berkeley, CA, 1973)
Franits, Wayne, *Dutch Seventeenth-century Genre Painting: Its Stylistic and Thematic Evolution* (New Haven, CT, and London, 2004)
Frierson, Cathy, *All Russia is Burning: A Cultural History of Fire and Arson in Late Imperial Russia* (Seattle, WA, and London, 2002)
Frost, Robert, *Robert Frost: Collected Poems, Prose, and Plays*, ed. Richard Poirier and Mark Richardson (New York, 1995)
Gagliadi, Jason, 'Behind the Secret of the Naga's Fire', www.time.com/magazine/article/0,9171.391567,00.html
Gilbert, Martin, *Second World War* [1989] (London, 2000)
Giltaij, Jeroen, et al., *Senses and Sins: Dutch Painters of Daily Life in the Seventeenth Century*, exh. cat., Museum Boijmans Van Beuningen, Rotterdam, and Städelsches Kunstinstitut und Städtische Galerie, Frankfurt am Main (Ostfildern, 2005)

Glen, Heather, ed., *The Cambridge Companion to the Brontës* (Cambridge, 2002)
Goldberg, Roselee, *Performa: New Visual Art Performance* (New York, 2005)
Gray, Basil, ed., *The Arts of India* (Oxford, 1981)
'The Great Fire', *St Louis Times*, 11 October 1871, p. 1
The Great Fire: Leading Newspaper Accounts of the Terrible Chicago Conflagration (St Louis, MO, 1871)
Griebel, Otto, *Ich war ein Mann Strasse: Lebenserinnerungen* (Altenburg, 1995)
Grimal, Pierre, *A Concise Dictionary of Classical Mythology* [1951], ed. Stephen Kershaw, trans. A. R. Maxwell-Hyslop (Oxford and Cambridge, MA, 1986)
Groys, Boris, in *Martin Honert*, exh. cat., Matthew Marks Gallery, New York (2002)
——, 'Mind's Eye Views: Boris Groys Talks with Martin Honert', *Artforum*, XXXIII/6 (February 1995)
Guest, Anthony-Haden, 'In Defense of Burning Man', *Art in America* (June/July 2005), p. 114
Hanna, Michelle, *Reconciliation in Olympism: Indigenous Culture In the Sydney Olympiad* (Petersham, New South Wales, 1999)
Hanson, Neil, *The Dreadful Judgement: The True Story of the Great Fire of London* (London and Sydney, 2001)
Harris, Ann Sutherland, *Seventeenth-Century Art and Architecture* (London, 2005)
Harth, Marjorie L., ed., *José Clemente Orozco: Prometheus* (Los Angeles and Santa Barbara, CA, 2001)
Healy, John, 'From Beyond the Grave', *The Mayo News* (15 November 2006), www.mayonews.ie/index.php?Itemid=35&id=620
'Hell's Fury', *Daily Telegraph*, 9 February 2009, p. 1
Herrmann, Wolfgang, *Gottfried Semper: In Search of Architecture* (Cambridge, MA, and London, 1984)
'High Jump', www.rooneydesign.com/JackBeNimble.html
'The History of Matches', http://inventors.about.com/library/inventors/blmatch.htm
Hoff, Ursula, *The Art of Arthur Boyd* (London, 1986)
Hoffmann, Heinrich, 'The Dreadful Story of Harriet and the Matches', *Struwwelpeter: Merry Stories and Funny Pictures* [German, 1844; English, 1848](New York, n.d.), www.gutenberg.org/files/12116/12116-h/12116-h.htm
Holloway, Sally, *Courage High! A History of Firefighting in London* (London, 1992)
Hopkins, David, *Dada and Surrealism: A Very Short Introduction* (Oxford, 2004)
Hopkinson, Tom, *The Blitz: London Then and Now* (London and Sydney, 1990)
Hountou, Julian, 'Le corps au mur: La méthode photographique de Gina Pane', *Etudes Photographiques*, no. 8 (November 2000), http://etudesphotographiques.revues.org/index229
Howatson, M. C., ed., *The Oxford Companion to Classical Literature*, 2nd edn (Oxford and New York, 1989)
Hultén, Pontus, *Jean Tinguely: A Magic Stronger than Death* (New York, 1987)
The Illustrated Gaelic-English Dictionary, ed. Edward Dwelly [1901–1911] (Glasgow, 1973)
Ingham, H. S., ed., *Fire and Water: An Anthology by Members of the NFS* [1942] (Exeter, 1992)
Ingram, Patricia, *The Brontës* (Oxford, 2006)
Irving, David, *The Destruction of Dresden* (London, 1963)

Itzkoff, Dave, 'The Rite of Summer', *New York Times Book Review* (19 September 2004), p. 14
'Jack be Nimble', www.rhymes.org.uk/jack_be_nimble.htm
James, Clive, 'Fires Burning, Fires Burning', *Weekend Australian* (23–4 September 2006), p. 3
'John French – The Art of Distillation', book five of six (London, 1651), www.levity.com/alchemy/jfren_ar.html
Johnson, Vivien, *Once Upon a Time in Papunya* (Melbourne, 2010)
Judd, Craig, and Amanda Lawson, 'Introduction', in *Euan Macleod, Napoleon Reef: Paintings 1992–2002*, exh. cat., Bathurst Regional Art Gallery, Bathurst (2002)
Juers, Evelyn, 'Janis Joplin and The Sisters of Excess', *Heat 7* (Sydney, 1998)
Kemp-Welch, Klara, 'Jerzy Bereś's Radical Expansion of the Readymade Event', unpublished draft, n.d.
Kitiarsa, Pattana, 'Faith and Films: Countering the Crisis of Thai Buddhism from Below', *Asian Journal of Social Sciences*, XXXIV/2 (2009)
Krell, Alan, 'In its Element', in *Fireworks: Tracing the Incendiary in Australian Art*, ed. Gavin Wilson, exh. cat., Artspace Mackay, Mackay City Council (Queensland, 2005), pp. 52–6
Landy, Michael, *Man in Oxford Street is Auto-Destructive*, exh. cat., Sherman Galleries, Sydney (2007)
Latham, Robert, and William Matthews, eds, *The Diary of Samuel Pepys* (London, 1972)
Lingwood, James, ed., *Susan Hiller: Recall: Selected Works 1969–2004* (Gateshead, 2004)
Lippard, Lucy, *Six Years: The Dematerialization of the Art Object* (New York, 1973)
Little, Tom, 'Las Fallas Festival', *Out of Town*, www.barcelonametropolitan.com/Article.aspx?TabID=2&MenuID=3&
Lowe, David, ed., *The Great Chicago Fire in Eyewitness Accounts and 70 Contemporary Photographs and Illustrations* (New York, 1979)
Lumby, Catharine, *Tim Storrier: The Art of the Outsider; Dust and Ashes*, exh. cat., Sherman Galleries, Sydney (2000)
MacCulloch, Alexandra, Keeper of Art, Clothing and Textiles, Buckinghamshire County Museum, correspondence with the author, 4 July 2006
McCutcheon, Russell T., *Manufacturing Religion: The Discourse on Sui Generis Religion and the Politics of Nostalgia* (Oxford and New York, 1997)
McKee, Alexander, *Dresden 1945: The Devil's Tinderbox* (London, 1982)
Maggi, Dr Pietro, Uster city archivist, correspondence with the author (2 August 2007)
Martin, Rupert, *The Sculpted Forest: Sculptures in the Forest of Dean* (Bristol, 1990)
Mayhew, Henry, *Mayhew's London*, ed. Peter Quennell (London, 1984)
Minogue, Sally, 'Introduction', in Charlotte Brontë, *Jane Eyre* (Ware, 1999)
Mitter, Partha, *Indian Art* (Oxford, 2001)
'The Monster Arrives Early', *Sydney Morning Herald*, 9 October 2002, p. 1
Murphy, Howard, *Aboriginal Art* (London, 1998)
'Naga Fireballs Festival 28 Oct 2005', www.discoverythailand.com/naga_fireballs_festival_2005.asp
National Arts Collections Fund, 'Review', National Gallery of Scotland (Scotland, 1990)
Neville, John, 'Introduction', in Tom Hopkinson, *The Blitz: London Then and Now* (London and Sydney, 1990)

Nicholls, Christine, ed., *The Magic Fire at Warlukurlangu: A Dreaming Narrative Belonging to Dolly Daniels Granites Nampijinpa* (Adelaide, 2003)
Noever, Peter, and F. Perrin, eds, *Yves Klein: Air Architecture* (Los Angeles, 2004)
Ntabeni, Mphuthumi, 'Burning Issues', *G21 Africa #438, Tenth Anniversary Special Edition 1996–2006*, http://g21.net/africa118.html
'Obituary: Paul Dessau', *Daily Telegraph*, 25 September 1999
'Obituary: Paul Dessau', *The Times*, 23 October 1999
O'Dell, Kathy, *Contract with the Skin: Masochism, Performance Art, and the 1970s* (Minneapolis, MN, and London, 1998)
O'Doherty, Brian, *Inside the White Cube: The Ideology of the Gallery Space* (London, 1976)
'The Olympic Flame and Torch Relay', www.hickoksports.com/history/olflame/shtml
Opie, Iona, and Peter Opie, *The Oxford Dictionary of Nursery Rhymes* [1951] (Oxford, 1997)
O'Riley, Michael Kampen, *Art Beyond the West: The Arts of Africa, India and Southwest Asia, China, Japan and Korea, The Pacific, and the Americas* (London, 2001)
Ossian, Rob, 'Pirates, Privateers, Corsairs, Explorers & More', www.geocities.com/Athens/7012/bios/index.htn
Parini, Jay, *Robert Frost: A Life* (London, 2001)
Parr, Mike, 'Some Notes On My Work 1970–1974', in *Mike Parr Performances 1971–2008* (Melbourne, 2008)
Pearce, Barry, *et al.*, *Arthur Boyd: Retrospective*, exh. cat., Art Gallery of New South Wales, National Gallery of Victoria, Tasmanian Museum and Art Gallery and Art Gallery of Western Australia (Sydney, 1994)
Pedlow, David, letter to *The Guardian*, 14 February 2004
Perkins, Hetti, and Hannah Fink, eds, *Papunya Tula: Genesis and Genius*, exh. cat., Art Gallery of New South Wales, Sydney, 2000)
Perlein, Gilbert, and Bruno Corra, *Yves Klein: Long Live the Immaterial*, exh. cat., Musée d'Art Moderne et d'Art Contemporain, Nice, and Museo Pecci, Prato (New York, 2001)
Peters, Harry T., *Currier & Ives: Printmakers to the American People* (New York, 1942)
Philips, Graham, *The Moses Legacy: In Search of the Origins of God* (London, Basingstoke and Oxford, 2002)
Phillips, Nicky, 'Singed but Safe: Test House Survives the Ravages of "Extreme"' Bushfire', *Sydney Morning Herald*, 17–18 April 2010, p. 13
Pimlott, Mark, 'Human World', *Frieze* (March–April 1994)
Piotrowski, Piotr, *In the Shadow of Yalta: Art and the Avant-Garde in Eastern Europe, 1945–1989*, trans. Anna Brzyski (London, 2009)
Porter, Stephen, *The Great Fire of London* (London, 1996)
Pyne, Stephen J., *Burning Bush: A Fire History of Australia* (New York, 1991)
Reid, Alexander, *Aye Ready: The History of the Edinburgh Fire Brigade, The Oldest Municipal Brigade in Britain* (Edinburgh, n.d.)
Reilly, Bernard, Jr, *Currier & Ives: A Catalogue Raisonné*, 2 vols (Detroit, MI, 1984)
Restany, Pierre, *Yves Klein: Fire at the Heart of the Void* (Putnam, 2005)
Richards, Mary, *Ed Ruscha* (London, 2008)
Roberts, Chris, *Heavy Words Lightly Thrown: The Reason Behind the Rhyme* (London, 2003)

Roberts, Ainslie (paintings), and Dale Roberts (text), *Shadows in the Mist: Australian Aboriginal Myths in Paintings* (Blackwood, South Australia, 1989)
Rutledge, Virginia, 'Suddenly It All Makes Sense', *Art in America* (June 1997)
Saltzer, Jerry, 'Notes on a Film', *Arts Magazine* (April 1988)
Sansom, William, James Gordon, et al., *Jim Brady: The Story of Britain's Firemen* (London, 1943)
Schjeldahl, Peter, 'A Woman's Work', *The New Yorker* (29 June 2009)
Scherle, Brian, 'The Heart is the Object of Fire: Fire and Passion in *Jane Eyre*', www.umd.umich.edu/casl/hum/eng/calsses/434/carweb/FIRE.htm
Schröder, Klaus Albrecht, and Johann Winkler, eds, *Oskar Kokoschka*, trans. David Britt (Munich, 1991)
Schwabsky, Barry, and Calvin Tomkins, *Michael Landy H2NY* (New York, 2007)
Science Museum, London, www.ingenious.org.uk/See/?target=SeeLarge&ObjectID={6...
Science Museum, London, www.ingenious.org.uk/See/?target=SeeLarge&ObjectID={7
Scorr, David W., '*Prometheus* Revisited', in *José Clemente Orozco: Prometheus*, ed. Marjorie L. Harth (Los Angeles and Santa Barbara, CA, 2001), pp. 27–45
Selden, Mark, 'A Forgotten Holocaust: US Bombing Strategy, the Destruction of Japanese Cities and the American Way of War from World War II to Iraq', www.japanfocus.org/-Mark-Selden/2414
'Self-Immolation in Vietnam', www.francesfarmersrevenge.com/stuff/archive/oldnews2/sel
Semper, Gottfried, *The Four Elements of Architecture and Other Writings*, trans. Harry Francis Malgrave and Wolfgang Herrmann (Cambridge, 1989)
Slive, Seymour, *Dutch Painting 1600–1800* (New Haven, CT, and London, 1995)
Smith, Carl, *Urban Disorder and the Shape of Belief: The Great Chicago Fire, the Haymarket Bomb, and the Model Town of Pullman* (Chicago and London, 1995)
Sotheby's Australia, *Fine Australian Paintings*, sale cat. (Sydney, 22 November 1992)
Stich, Sidra, *Yves Klein* (Ostfildern, 1994)
Sutton, Peter, ed., *Dreamings: The Art of Aboriginal Australia*, exh. cat., New York, Chicago, Melbourne and Adelaide (1988/9)
Sylvester, David, ed., *René Magritte: Catalogue Raissoné*, 5 vols (London, 1993)
Taylor, Frederick, *Dresden: Tuesday 13 February 1945* (London, 2004)
Tinguley, Jean, et al., *Jean Tinguely Retrospective* (Boston, 2008)
Tinguely, Jean, and Pontus Hultén, *Jean Tinguely: Meta* (London and Boston, 1975)
Tinniswood, Adrian, *By Permission of Heaven: The Story of the Great Fire of London* (London, 2003)
Torczyner, Harry, *Magritte: Ideas and Images*, trans. Richard Miller (New York, 1979)
Trazaskoma, Stephen M., et al., eds, *Anthology of Classical Myth: Primary Sources in Translation* (Indianapolis, IN, 2004)
Tyrrell, Peter, *Founded on Fear* (Dublin, 2006)
Viso, Olga M., *Ana Mendieta. Earth Body: Sculpture and Performance, 1972–1985* (Ostfildern-Ruit, 2004)
——, *Unseen Mendieta: The Unpublished Works of Ana Mendieta* (Munich, Berlin, London and New York, 2008)
Vitruvius, *Ten Books on Architecture*, trans. Ingrid D. Rowland (Cambridge, 1999)

Vonnegut, Kurt, Jr, *Slaughterhouse-Five* [1969] (London, 2000)
Wainright, Robert, 'More Spin than a Warnie Wrong "un"', *Sydney Morning Herald*, 18–19 July 2009, p. 3
Walker, John A., *Firefighters in Art and Media: A Pictorial History* (London, 2005)
Wallace, Karen, *London's Burning!* (London, 1997)
Wallis, Simon, 'Fire Fiesta', http://homepage.mac.com/simon.wallis/site/newspapers/fallas.htm
Walsh, Bryan, 'Why Global Warning May Be Fueling Australia's Fires', *Time* (9 February 2009), www.time.com/time/health/article//0,8599,1878220,00.html
Walsh, Gary, 'Great Balls of Fire', *Sydney Morning Herald*, 15 November 2003
——, 'River of Fire', *Sydney Morning Herald*, 13–14 March 2004
Watson, Neil, *Burn: Artists Play With Fire*, exh. cat., Norton Museum of Art, West Palm Beach, FL (2001)
Weitemeier, Hannah, *Yves Klein, 1928–1962: International Klein Blue* (Cologne, 1995)
Whitford, Frank, *Oskar Kokoschka: A Life* (London, 1986)
Wilson, Gavin, *Fireworks: Tracing the Incendiary in Australian Art*, exh. cat., Artspace Mackay, Mackay City Council (Mackay, Queensland, 2005)
Wilton, Andrew, *Turner and the Sublime* (London, 1980)
Wright, Angela, *Leonard Rosoman: A War Retrospective, 1939–45*, exh. cat., Imperial War Museum, London, and the Fine Arts Society, Edinburgh (1989)
Wright, Thomas, *The Romance of the Lace Pillow* (Carlton, 1982)
Zelevansky, Lynne, 'An Interview with Ed Ruscha', in *Magritte and Contemporary Art: The Treachery of Images*, ed. Stephanie Barron and Michel Draguet with the assistance of Sara Cochran, exh. cat., Los Angeles County Museum of Art, Los Angeles (2007)
Zimmermann, Anja, '"Sorry for Having to Make You Suffer": Body, Spectator, and the Gaze in the Performances of Yves Klein, Gina Pane, and Orlan', *Discourse* XXIV/3 (2002), pp. 27–46, http://muse.jhu.edu/journals/discourse/v024/24.3zimmermann.html

ACKNOWLEDGEMENTS

This book has had a long gestation. Many thanks to friends and colleagues who encouraged me to 'just get it done'. My editor, Michael Leaman, matched their patience. And to all others at Reaktion Books who helped in the realization of the project, thank you. The staff of libraries, museums and photo-agencies have been most helpful: I would like to thank especially Art Resource, New York; the Imperial War Museum, London; the Museum of Fire, located in the Lothian and Borders Fire Brigade HQ, Edinburgh; the Chicago Historical Society; the Africa-American Firefighter Museum, Los Angeles; the library of the Art Gallery of New South Wales, Sydney, and the Clement Semmler Library at the College of Fine Arts (COFA), The University of New South Wales. To COFA, I am grateful for a sabbatical and financial support. Staff in Print and Digital Media were most helpful, while Graham Forsyth and Martin Sims provided valuable criticism, and Doug McKeogh was always there with a good wine. For a room to stay and convivial times, I thank my very dear friends in New York, England and Amsterdam: Rigbie Turner; Colin and Melanie Raizon and Pam and Philip Wood; Peter and Jeannine Herzstein. *Merci, aussi,* Dr Wayne Mason. My family as always has been a source of great support. To all institutions and individuals who granted copyright permission, and to those who generously waived fees, I am most appreciative. A special thanks must go to Sheila Christofides who, in my last year of writing, was a tower of strength (here the cliché is spot on): she worked assiduously on the demanding tasks of referencing, bibliography and copyright; her comments on style and content were also much welcomed, as were her sandwiches. I promised my late partner, Andrea Stretton – journalist, interviewer and arts presenter – who died on 16 November 2007, that this book would be dedicated to her and her darling granddaughter, the aptly named Phoenix Lila Southall. It is.

PHOTO ACKNOWLEDGEMENTS

The author and publishers wish to express their thanks to the below sources of illustrative material and/or permission to reproduce it:

Photo AAP/Lacy Atkins, © 2004 AAP: p. 91; photo AAP/Malcolm Browne © 1963 AAP: p. 84; photos © ADAGP, Paris and DACS, London 2011: pp. 156, 157, 160, 175; © Suzanne L. Adams, courtesy of the artist: p. 94; Art Gallery of New South Wales, Sydney (gift of Ms Catherine Palmer 1983 – © and reproduced with permission of the Bundanon Trust, North Nowra, NSW): p. 80; Art Gallery of South Australia, Adelaide (gift of Diana Ramsay AO and the Art Gallery of South Australia Foundation in celebration of the Foundation's 20th Anniversary 2001): p. 72; Ballarat Fine Art Gallery (gift of Lady Currie in memory of her husband, the late Sir Alan Currie, 1948): p. 61; reproduced courtesy bpk, Berlin: p. 36; photo br0, courtesy of stock.xchng: p. 6; Bundanon Collection, North Nowra, NSW (© and reproduced with the permission of the Bundanon Trust): p. 79; photo Pierre Joly-Véra Cardot, © Pierre Joly-Véra Cardot: p. 169; photo Central Street Gallery, Sydney, Australia: p. 178; Centre d'essai de Gaz de France: pp. 168 left (photo no. YK ph_0975, classification no. 19610232, Yves Klein réalisant F2), 168 right (photo no. YK ph_0989, classification no. 19610232, Yves Klein réalisant F2), 169 (photo no. YK ph_0992, classification no. 19610232, Yves Klein réalisant FC 1); photo courtesy Chicago History Museum: pp. 42, 43, 44; Dillon Cohen collection: p. 171; The Samuel Courtauld Trust, Courtauld Institute of Art Gallery, London: p. 69; courtesy of the artist (Fred Cress): p. 10; © 2001 DVD Productions Pty Ltd: p. 189; used by permission of Fairfaxphotos/John Fairfax Publications: pp. 186, 187, 188; courtesy of the Fire-Works Gallery, © Ronnie Tjampitjimpa, the artist licenced by Aboriginal Artists Agency 2007: p. 60; © Peter Fischli and David Weiss, courtesy Matthew Marks Gallery, New York: pp. 139, 140, 141; © Shane Fitzgerald, reproduced by kind permission of the artist: p. 62; © Fondation Oskar Kokoschka/DACS 2011: p. 69; photo Louis Frédéric, © Louis Frédéric: p. 168 (right); courtesy of the artist (Philip Gostelow): p. 16; Klaus Groenke Collection, Berlin: p. 157; from *Harper's Weekly* (28 October 1871): p. 44; Hirshhorn Museum and Sculpture Garden, Smithsonian Institution, Washington, DC: p. 155; Martin Honert, courtesy Matthew Marks Gallery, New York: p. 149; reproduced with permission of Imperial War Museum, London: pp. 27, 38; photo Inhibodress Gallery, Charles Street, Woolloomooloo, NSW, Australia: p. 177; The J. Paul Getty Museum, Los Angeles: pp. 8, 11, 77, 101; photo © Heino Kalis/Reuters/Picture Media: p. 86; courtesy the artist (Michael Landy): p. 148; courtesy the artist (Michael Landy), Thomas Dane Gallery, London and Sherman Galleries, Sydney: p. 147; London Fire Brigade Museum, © London Fire and Emergency Planning Authority: p. 22, 23; photo London Metropolitan Archives: p. 29; Los Angeles County Museum of Art (gift of The Ahmanson Foundation (M.82.7): p. 73 (digital image © 2009 Museum Associates/

LACMA/Art Resource NY); courtesy of the artist (Euan Macleod): p. 83; photo Andrew Meares (used by permission of Fairfaxphotos/John Fairfax Publications): p. 19; The Menil Collection, Houston, TX: p. 160; The Metropolitan Museum of Art, New York (purchase, Maria-Gaetana Matisse gift, 1993): p. 103 (image © The Metropolitan Museum of Art/Art Resource, NY); Nick Moir/Fairfaxphotos, used by permission of Fairfax photos/John Fairfax Publications: p. 96; Musées Royaux des Beaux-Arts de Belgique, Brussels (photo irpa-kik): p. 106; from Dolly Daniels Granites Nampijinpa, *The magic fire at Warlukurlangu: a Dreaming narrative belonging to Dolly Daniels Granites Nampijinpa* (Kingswood, S. Aust.: Working Title Press, 2003) – series edited by Christine Nicholls, assisted by Sue Williams; reproduced by permission of Cecily Napanangka Granites: p. 54; National Galleries of Scotland, Edinburgh: pp. 109 (purchased with the assistance of the Art Fund and the National Heritage Memorial Fund 1989), 110 (accepted in lieu of tax, with additional funding from the National Heritage Memorial Fund, the Art Fund and Gallery funds 1989); National Gallery of Victoria, Melbourne: pp. 25, 55; National Library of Australia, Canberra: p. 58 (top); courtesy of the artist (Cornelia Parker) and Frith Street Gallery, London: p. 14; © Mike Parr: pp. 177, 178; courtesy of the photographer Darren Pateman and John Fairfax Holdings: p. 20; Pomona College Museum of Art, Claremont, CA: p. 70; photos Stefan Postles: pp. 186, 187, 188; private collections: pp. 60, 62, 83, 104, 156; © the Dean and Chapter of St Paul's Cathedral, London: p. 34; Rijksmuseum, Amsterdam: pp. 67, 108; Royal Cabinet of Paintings 'Mauritshuis', The Hague: p. 111; © Ed Ruscha, courtesy Gagosian Gallery, New York: pp. 152, 153, 155; Science and Society Picture Library, Science Museum, London: pp. 114, 115, 116; courtesy of the artist (Wendy Sharpe): p. 104; Staatliches Museum Schwerin (photo Elke Walford): p. 105; photo Tony Stanyer Photography: p. 60; Tate, London: p. 184; photo Timothy Taylor Gallery, London: p. 184; custodian Tjapaltjarri-Tjungurrayi, © Dorn Bardon: pp. 57, 58 (foot); Walker Art Center, Minneapolis (Clinton and Delia Walker Acquisition Fund, 1993): p. 137; from Karen Wallace, *London's Burning* (London, 1997), first published by Franklin Watts, a division of The Watts Publishing Group Ltd, 338 Euston Road, London, NW1 3BH: p. 31; photo Bernward Wember: p. 166; photo Ray Woodbury: p. 80; photo by Yatzek (Jacek Szmuc): p. 181.

INDEX

Aboriginal Dreamings 52–64
Bungalung Old Man's Fire Story (Tjupurrula) *26*, 58, 59
Bushfire between Mount Elephant and Timboon (Von Guérard) *28*, 59–61
Bushfire (Tjampitjinpa) *27*, 59, 60
ceremonials and fire 62–4
firestick farming 56, 61–2
'Fishing by Torchlight' (Lycett) *25*, 56–9
The Magic Fire of Warlukurlangu (Daniels) *22*, 53–4
and symbolism of fire 52–3
Warlukurlangu Fire Dreaming *22–3*, 53–6
Warlukurlangu Jukurrpa (Fire Country Dreaming) (Danielsand Martin) *23*, 55, 56
Woman's Fire Dreaming (Tjapaltjarri) *24*, 56, 57
Abramović, Marina
Rhythm O 172
'Seven Easy Pieces' 174
Adams, Suzanne, *Fire Talk* *42*, 94, 95–6, 97
Aeschylus, *Prometheus* trilogy 65–6
Alberti, Leon Battista, *On the Art of Building* 39, 99–100
Ancient Greece and Promethean condition *see* Prometheus
Andersen, Hans Christian, *The Little Match-Seller* 121–2
Anthropometries (Klein) 162, 165, 168
Antonioni, Michelangelo, *Zabriskie Point* 146
'The Ashes' (Pateman) *7*, 19–20
Australia
Aboriginal Dreamings *see* Aboriginal Dreamings

Black Saturday bushfires, Victoria (2009) 18–19
bushfires, buildings able to withstand *81–3*, 186, 187–8
Sydney bushfires (2002) *5–7*, 16, 18, 19–21
Sydney, lighting of Paralympic flame 63–4
Umina fires, New South Wales *43*, 96–7
Autoportrait(s) (Pane) *76*, 173, 174, 175
Baburen, Dirck van, *Prometheus being Chained by Vulcan* *30*, 66–7
Bachelard, Gaston
Fragments of a Poetics of Fire 65
The Psychoanalysis of Fire 100, 127, 161–2, 163
Backdraft (film) 24
Banham, Reyner 97
Bealtine festival 123
Belshazzar's Feast (Hiller) *80*, 183–4
Bening, Simon, 'Moses and the Burning Bush' *35*, 76, 77
Bereś, Jerzy
Krzysztofory Gallery, Kraków performance *79*, 179–80, 181
Prophesy II 179
Romantic Manifestation 179, 180–82
Berlin, book burning *18*, 37–8
Best, David, 'Temple of Stars' 92
Beuys, Joseph 161
Bhalo, Ahmed Nassir Juma, 'The Fire' 135
Blake, William, *Nebuchadnezzar* 78

The Blitz: London's Firestorm (film) 28
blue colour and fire 162–3, 164
Feu de Bengale (Bengal Flares) (Klein) 162, 163
Boase, T.S.R., *Nebuchadnezzar* 78, 79
bodies, fire's physical effects on 51–2
and performance art *see* performance art
senses, fire's effect on 11–12
book burning, Berlin *18*, 37–8
Booth, William and Catherine 117
Boyd, Arthur
The Fiery Furnace 78
Love, Marriage and Death of a Half-caste 82
Nebuchadnezzar in a Fire 36, 78, 79
Nebuchadnezzar on Fire Falling over a Waterfall 37, 78–80
Red Nebuchadnezzar with Crows 79, 81–2
Boyles, Robert 112, 113–14
Braidwood, James 48–9
Brisley, Stuart, *Being and Doing* 179, 180–82
Broder, Steve, 'Books are Weapons on the War of Ideas' *18*, 38
Brontë, Charlotte, *Jane Eyre* 127–32, 133
candle references 130, 131, 134
Bryant & May 120–21
Burden, *Prelude to 220, or 110* 172
Burn: Artists Play With Fire exhibition 185
Burning Bush (Old Testament) *35*, 76–8
'burning glasses' *53*, 113, 114
The Burning Landscape (Magritte) 158
Burning Man Festival, US *41*, 51, 89–93
bushfires
Black Saturday bushfires, Victoria (2009) 18–19
buildings able to withstand, Australia *81–3*, 186, 187–8
Bushfire between Mount Elephant and Timboon (Von Guérard) *28*, 59–61
Bushfire (Tjampitjinpa) *27*, 59, 60
Sydney (2002) *5–7*, 16, 18, 19–21

candles
in *An Allegory* (El Greco) 107–10
in *Autoportraits* (Pane) 174
in *Boy Blowing on a Firebrand to Light a Candle* (Schalcken) *50*, 107, 109
candlestick dance 126–7
candlestick, jumping over *see* 'Jack Be Nimble'
in *Der Lauf Der Dinge* (Fischli and Weiss) 138, 139
in *La Lampe Philosophique* (Magritte) *68*, 156
in *Light a Candle* (Parr) *77*, 176, 177
menorah and Hannukah 90
in *The Night School* (Dou) *49*, 107, 108
Nocturnal Day of the Dead, Mexico 169
in *The Proposition* (Leyster) *52*, 110–12
references to, in *Jane Eyre* 130, 131, 134
in *Silueta* (Mendieta) 139, 170
in *Soldiers around a Fire* (Duyster) *47*, 102–4, 105
and 'study' metaphor 107
see also small-scale fires
Chamberlin, Everett, *Chicago and the Great Conflagration* 40–41
Chamberlin, Joseph Edgar 44–5
Chapin, John R., *The Rush for Life over Randolph Street Bridge* *21*, 42–4
Chicago, Great Fire of 17, *19–21*, 39–45
Chicago in Flames (Flint) *19*, 41, 42
Chicago and the Great Conflagration (Colbert and Chamberlin) 40–41
Destruction of Chicago by Fire (Kelly) *20*, 41–2, 43
children
'The Dreadful Story of Harriet . . .' (Hoffman) *57*, 117–20
lacemakers and child labour 125–6
play, and fear and fun 136
The Little Match-Seller (Andersen) 121–2
Colbert, Elias, *Chicago and the Great Conflagration* 40–41
Congreve, William 114–15
Coomaraswamy, Ananda K., *The Dance of Siva* 74–5
Costa, Alex 167
Cress, Fred, *Two Waves* *2*, 9, 10

Currier, Nathaniel, 'The Life of a Fireman' 24

Damage (Ruscha) *65*, 151, 152
Daniels, Dolly Granites Nampijinpa 53
  The Magic Fire of Warlukurlangu *22*, 53–4
  Warlukurlangu Jukurrpa (Fire Country Dreaming) *23*, 55, 56
Dante, *Inferno* 71
Denning, Mark, DVD *Fireplace* *84*, 189–90
Dessau, Paul, *Menace* *8–11*, 21–3
Dibbet, Jan, *TV as a Fireplace* 182
The Discovery of Fire (Magritte) 158
Domenichino, *Landscape with Moses and the Burning Bush* 76
Dou, Gerrit
  The Night School *49*, 107, 108
  Young Man Drawing after a Sculpture *48*, 104–7
Dresden bombing, WWII 17–18, 35–8, 51
  View of the Dresden Ruins... (Hahn) *17*, 36
Druids, May-day festival 123
Duyster, Willem, *Soldiers Around a Fire* *47*, 102–4, 105
Dwelly, Edward 123

Edinburgh, Great Fire of 17, 46–50
Edison, Thomas, *Life of An American Fireman* 24
El Greco, *An Allegory* 107–10
emotions, conflicting *see* fire associations, conflicting
Evelyn, John 28, 34
The Eye of the Mountain (Magritte) 158

Fernández-Galiano, Luis, *Fire and Memory* 97
fertility and fire jumping 123
festive fires 85–93
  Bealtine festival 123
  Burning Man Festival, US *41*, 51, 89–93
  Druids, May-day festival 123
  Las Fallas festival, Spain *40*, 85–7
  Naga Fireballs of Nong Khai, Thailand 88–9
Feu de Bengale (Bengal Flares) (Klein) 162, 163
Feuer (Fire) (Honert) *64*, 147–8, 149, 150–51
The Fiery Furnace (Boyd) 78
fire associations, conflicting
  allure and ruin 120
  awe and horror 45
  cleansing and renewal 85–6, 93
  'deadly prankster' 136, 145, 185–90
  destruction and redemption 7, 13, 52–3, 61–2, 74–5, 82, 170, 179
  disfigurement and warmth 51–2
  drama and intensity 165–6
  enchantment 95–7
  fascination and fear 17, 141
  fear and fun 136
  fire and pain 173, 176–8
  fire and water *see* fire and water
  good and evil 82, 162
  heavenly judgement, fire seen as 47
  Hell and damnation *34*, 71–5
  marvellous and dreadful 40
  redemption and purification 141
  sublime, as description 17, 40, 43, 45, 47, 61
  and vocabulary of fire 11, 12–13, 18, 128, 163
  see also symbolism
'The Fire' (Bhalo) 135
Fire Column (Klein) *71*, 165, 166
fire creation, 'burning glasses' 53, 113, 114
fire, discovery of 98–9
The Discovery of Fire (Magritte) 158
'Fire and Ice' (Frost) 132–3
fire jumping 122–7
  Bealtine festival 123
  candlestick dance 126–7
  Chaharshanbeth-suri festival, Iran 126
  and fertility 123
  and good luck 123
  in 'Jack Be Nimble' 122–3
  and lacemakers 124–5
  and St Catherine's Day 124
  and weddings 124
Fire and Memory (Fernández-Galiano) 97
'fire paintings' (Klein) *72–4*, 165–7, 168–9
Fire Talk (Adams) *42*, 94, 95–6, 97

The Fire of Uster (Fischli and Weiss) *58*, 137–8
Fire Wall (Klein) 165
fire and water
  Architecture of Air (Klein) 163–5
  'Fire and Ice' (Frost) 132–3
  Fire and Water, Sansom, William 11–12, 26–8
  Two Waves (Cress) *2*, *9*, *10*
  see also fire associations, conflicting
'Fire-bottles' (Magritte) *70*, 159–60
fireballs, Naga Fireballs of Nong Khai, Thailand 88–9
firebird *see* Phoenix
firedogs *44*, 101, 113
  see also hearth and home
firefighters, heroism of 23–8
firemen artists *8–11*, 21–3
fireplaces, first appearance of 100–101
  see also hearth and home
'Fires Burning, Fires Burning' (poem) (James) 49–50
fireworks 29, 85, 138, 169
  Naga Fireballs of Nong Khai, Thailand 88–9
  Silhouette of Fireworks (Mendieta) 170
Fischli, Peter
  Der Lauf Der Dinge (The Way Things Go) *59*, *60*, *61*, 136, 138–42, 143
  The Fire of Uster *58*, 137–8
  'Wurstserie' ('Sausage Series') 136–7
Fitzgerald, Shane, *Inferno, Arnhem Land* *29*, *60*, *61*, *62*
Flint, William, *Chicago in Flames* *19*, *41*, *42*
Food, TV News, Fire (Pane) 173–4
forest fires
  'The Fire' (Bhalo) 135
  see also bushfires
French, John, *The Art of Alchemy* 112–13
Frost, Robert, 'Fire and Ice' 132–3

gendering of fire 24
Germany
  Berlin, book burning *18*, 37–8
  Dresden *see* Dresden bombing, WWII
global warming 186–7
Gostelow, Philip, *Backyard, Woodford* *5*, 16, 21
Greece, Athens political protests 185
Griebel, Otto 37
Griffier the Elder, Jan, *London on Fire* *14*, 29
guitar burning (Hendrix) 92

Hahn, Walter, *View of the Dresden Ruins...* *17*, 36
Hanging Fire (Parker) *4*, 13, 14
Harvey, Larry 89–90
hearth and home 97–101
  Belshazzar's Feast (Hiller) *80*, 183–4
  Black Saturday bushfires, Victoria (2009) 18–19
  bushfires, buildings able to withstand *81–3*, 186, 187–8
  DVD Fireplace (Denning) *84*, 189–90
  firedogs *44*, 101
  fireplaces, first appearance of 100–101
  Foto (Honert) 148–50
  Hearth (MacIver) *45*, 102, 103
  Hill End (Sharpe) *46*, 102, 104
  A Man Warming by a Fire (Vrelant) *1*, 7–9, 102
  On the Art of Building (Alberti) 39, 99–100
  reverie, fireplace as object of 100
  small fires 102–3
  Sydney bushfires (2002) *5–7*, 16, 18, 19–21
  Ten Books on Architecture (Vitruvius) 98–9
  Time Transfixed (Magritte) 156
  TV as a Fireplace (Dibbet) 182
  TV set as fireplace *80*, *84*, 183–4, 189–90
  Umina fires, New South Wales *43*, 96–7
  see also small-scale fires
heavenly judgement, fire seen as 47
Helios the sun god 65
Hell and damnation *34*, 71–5
Hendrix, Jimi, guitar burning 92
Hepple, Norman 22
Hiller, Susan, *Belshazzar's Feast* *80*, 183–4
Hinduism
  Rig Veda 75
  and Siva *33*, 71–5

Hoffman, Heinrich, *Struwwelpeter* and 'The Dreadful Story of Harriet . . .' *57*, 117–20
Homage (Maler) 170
Honert, Martin
Feuer (Fire) *64*, 147–8, 149, 150–1
Foto 148–50

Icarus 81
image and text, relationship between 151–5
see also Ruscha, Ed
Inferno, Arnhem Land (Fitzgerald) *29*, *60*, *61*, *62*
Inferno (Dante) 71
Iran, Chaharshanbeth-suri festival 126
Ives, James Merritt, 'The Life of a Fireman' *24*

Jack Be Nimble' (nursery rhyme) 122–3
see also fire jumping
James, Clive, 'Fires Burning, Fires Burning' (poem) 49–50
Jane Eyre see Brontë, Charlotte, *Jane Eyre*
Japanese tinder pistol *56*, 113, 116
Jones, Samuel 112, 116–17

Kelly, Thomas, *Destruction of Chicago by Fire* *20*, 41–2, 43
Klein, Yves 142, 161–2
Anthropometries 162, 165, 168
Architecture of Air 163–5
Feu de Bengale (Bengal Flares) 162, 163
Fire Column *71*, 165, 166
'fire paintings' *72–4*, 165–7, 168–9
Fire Wall 165
'fireman' hoax 167
'Immaterial' 162
'International Klein Blue' (IKB) 163
'Leap into the Void' 162, 167
Klüver, Billy 143, 144, 145
Kokoschka, Oskar, *Prometheus* *31*, 67–8, 69

La Lampe Philosophique (Magritte) *68*, 156
lacemakers
and candlestick jumping 124–5
and child labour 125–6
Ladder 49 (film) 24–6
Ladder of Fire (Magritte) *69*, 157–8
Landy, Michael
BreakDown 146
H2NY/Homage to New York *62*, *63*, 146–7, 148
Der Lauf Der Dinge (The Way Things Go) (Fischli and Weiss) *59*, *60*, *61*, 136, 138–42, 143
Lennon, Mike, *Brothers . . . On Holy Ground* (film) 26
Lesueur, Charles-Alexandre 62–3
Leyster, Judith, *The Proposition* *52*, 110–12
London Blitz *8–11*, 17–18, 21–3
A House Collapsing on Two Firemen... (Rosoman) *13*, 26–8
The Blitz: London's Firestorm (film) 28
St Paul's Cathedral *16*, 34
London, Great Fire of 13, 17, 28–34
London on Fire (Griffier the Elder) *14*, 29
London Verbrandt (London on Fire) (engraving) 30
London's Burning! (Wallace) *15*, 29–30, 31
rebuilding plans after 34
Lucifers (matches) 117, 120
see also matches
Lundström brothers 120
Lycett, Joseph, 'Fishing by Torchlight' *25*, 56–9

MacIver, Loren, *Hearth* *45*, 102, 103
Macleod, Euan, *Exquisite Corpse with Fire* *38*, 81, 83
Magritte, René 151
The Burning Landscape 158
The Discovery of Fire 158
The Eye of the Mountain 158
'Fire-bottles' *70*, 159–60
The Human Condition 158–9
La Lampe Philosophique *68*, 156
Ladder of Fire *69*, 157–8
Time Transfixed 156
The Treachery of Images 156–7
Making Full Moon Party (film) 88–9
Maler, Leopoldo, *Homage* 170
Martin, Uni Nampijinpa, *Warlukurlangu*

Jukurrpa (Fire Country Dreaming) 23, 55, 56
matches
Bryant & May 120–21
Have a burning match dropped on your bare chest (Parr) 78, 177–8
history of 112, 113–17
Lucifers 117, 120
and phosphorus toxicity 117–18
and popular culture 57, 117–22
Prometheans 116–17
safety matches 120–21
Struwwelpeter and 'The Dreadful Story of Harriet . . .' (Hoffman) 57, 117–20
The Little Match-Seller (Andersen) 121–2
Thresholds No. 1; With Fire and Water (Parr) 178–9
Matilda of Austria, Archduchess 120
Maurer, Louis 24
Mendieta, Ana 161, 167–9
Burial of Ñañigo 170
Silhouette of Fireworks 170
Siluetas 75, 139, 168, 169, 170–2
Mexico, Nocturnal Day of the Dead 169
Millais, John Everett
Ophelia 154–5
The Rescue 12, 24, 25
Moir, Nick, Umina fires' photograph 43, 96–7
Montano, Linda, *Mitchell's Death* 172
'Moses and the Burning Bush' (Bening) 35, 76, 77

Naga Fireballs of Nong Khai, Thailand 88–9
nationalism, and metaphor of fire 135
Nebuchadnezzar, King of Babylon 77–82, 189
Nebuchadnezzar (Blake) 78
Nebuchadnezzar (Boase) 78, 79
Nebuchadnezzar in a Fire (Boyd) 36, 78, 79
Nebuchadnezzar on Fire Falling over a Waterfall (Boyd) 37, 78–80
'necklacing', South Africa 139–41
Nouveau Réalisme 142–6, 161

oil light
and 'study' metaphor 107
see also candles
Olympics
lighting of Paralympic flame, Sydney 63–4
Torch Relay, history of 64
Oppenheim, Dennis, *Rocked Circle – Fear* 172
Orozco, José Clemente, *Prometheus* 32, 68–9, 70

Pane, Gina
Autoportrait(s) 76, 173, 174, 175
Food, TV News, Fire 173–4
Unanaesthetized Climb 172
Parker, Cornelia, *Hanging Fire* 4, 13, 14
Parr, Mike 174–7
Have a burning match dropped on your bare chest 78, 177–8
Hold your breath for as long as possible 176
Light a candle 77, 176, 177
Thresholds No. 1; With Fire and Water 178–9
Pateman, Darren, 'The Ashes' 7, 19–20
Pepys, Samuel 28–34
performance art
Being and Doing (Brisley) 179, 180–82
Bereś, Jerzy *see* Bereś, Jerzy
Klein, Yves *see* Klein, Yves
Landy, Michael *see* Landy, Michael
Nouveau Réalisme 142–6, 161
Pane, Gina *see* Pane, Gina
Parr, Mike *see* Parr, Mike
Prelude to 220, or 110 (Burden) 172
Phoenix 13, 86–7
design in St Paul's Cathedral 16, 34
Picano, Francesco Antonio, *St Michael Casting Satan into Hell* 34, 71, 73
Plato 158, 159
Poland, artistic performance in protest 179–82
Prometheans (matches) 116–17
see also matches
Prometheus
and 'giving of fire' 68, 109
myth of 52, 65–70, 179, 186, 188–9

Promethean capacity, and creativity 179
Prometheus being Chained by Vulcan (Baburen) *30*, 66–7
Prometheus (Kokoschka) *31*, 67–8, 69
Prometheus (Orozco) *32*, 68–9, 70
Prometheus trilogy (Aeschylus) 65–6
The Psychoanalysis of Fire (Bachelard) 100, 127, 161–2, 163

Quang Duc, Thich 82–4

Rauschenberg, Robert 144
Red Nebuchadnezzar with Crows (Boyd) 79, 81–2
The Rescue (Millais) *12*, 24, 25
Rosoman, Leonard 22
A House Collapsing on Two Firemen . . . 13, 26–8
Ruhnau, Werner, Architecture of Air 163–5
Ruscha, Ed
Damage 65, 151, 152
Los Angeles County Museum on Fire 67, 154–5
Various Small Fires 66, 152–4
Russia, Burning Bush in Orthodox Church 76–7

safety matches 120–1
see also matches
St Catherine's Day, and fire jumping 124
Sansom, William, *Fire and Water* 11–12, 26–8
Sauria, Charles 117
Schalcken, Godfried, *Boy Blowing on a Firebrand to Light a Candle 50*, 107, 109
Schmidt, Julius 169, 170
self-immolation *39*, 52, 82–5
Semper, Gottfried, Dresden Synagogue 37, 38–9
senses, fire's effect on 11–12
see also bodies, fire's physical effects on
September 11, 2001 26, 185
sexualized fire, and *Jane Eyre* 127–34
Shadrach, Meshach and Abed-nego (Turner) 77–8
Sharpe, Wendy, *Hill End 46*, 102, 104
Siluetas (Mendieta) *75*, 139, 168, 169, 170–2
Siva Nataraja (Dancing Siva) 33, 71–4
Slaughterhouse-Five, Vonnegut, Kurt 37
small-scale fires 102–12
candles *see* candles
oil light, and 'study' metaphor 107
Various Small Fires (Ruscha) *66*, 152–4
see also hearth and home
snow, fire described as red 46
South Africa
'necklacing' 139–41
township burnings 185
Spain, *Las Fallas* festival *40*, 85–7
study and learning aided by fire 107
symbolism 170, 171–2
and *Jane Eyre* 127–32, 133
see also fire associations, conflicting

'Temple of Stars' (Best) 92
Thailand
anti-government conflict 185
Naga Fireballs of Nong Khai 88–9
The Expulsion of Paradise 3, 9–11
Time Transfixed (Magritte) 156
tinder box *54*, 113, 115
tinder pistol *55–6*, 113, 115, 116
Tinguely, Jean
Homage to New York 142–6
Landy's homage to 146–7
Tjampitjinpa, Ronnie, *Bushfire 27*, 59, 60
Tjapaltjarri, David Corby, *Woman's Fire Dreaming 24*, 56, 57
Tjupurrula, Johnny Warangkula, *Bungalung Old Man's Fire Story 26*, 58, 59
Tokyo, WWII bombing 49
The Tomahawk (journal) 120–21
Turner, J.M.W., *Shadrach, Meshach and Abed-nego* 77–8
Two Waves (Cress) *2*, 9, 10
Tyrrell, Peter 82

UK
Edinburgh, Great Fire of *17*, 46–50
London *see* London
US
Burning Man Festival *41*, 51, 89–93
Chicago *see* Chicago, Great Fire of

H2NY/Homage to New York (Landy) 62, 63, 146–7, 148
Homage to New York (Tinguely) 142–6
Los Angeles County Museum on Fire (Ruscha) 67, 154–5
Los Angeles wildfires 9, 13
Monterey Pop Festival, Hendrix guitar burning 92
New York and 9/11 26, 185
September 11, 2001 26, 185
volunteer fire departments 23–4

Vaccaro, Lorenzo, *St Michael Casting Sataninto Hell* 34, 71, 73
Various Small Fires (Ruscha) 66, 152–4
Vietnam War
artistic performances in protest 172–4
and napalm 82
and self-immolation 39, 82–5
Vitruvius, *Ten Books on Architecture* 98–9
vocabulary of fire 11, 12–13, 18, 128, 163
see also fire associations, conflicting
Von Guérard, Eugene, *Bushfire between Mount Elephant and Timboon* 28, 59–61
Vonnegut, Kurt, *Slaughterhouse-Five* 37
Vrelant, William, *A Man Warming by a Fire* 1, 7–9, 102

Walker, John 112, 114, 116, 117
Wallace, Karen, *London's Burning!* 15, 29–30, 31
wartime *see* Germany; London Blitz; Tokyo; Vietnam War
weddings, and fire jumping 124
Weiss, David
Der Lauf Der Dinge (The Way Things Go) 59, 60, 61, 136, 138–42, 143
The Fire of Uster 58, 137–8
'Wurstserie' ('Sausage Series') 136–7
The Wicker Man (film) 90, 123
wildfires
and global warming 186–7
Los Angeles 9, 13
see also bushfires; forest fires
witch burning 140
Woman's Fire Dreaming (Tjapaltjarri) 24, 56, 57

Zabriskie Point (Antonioni) 146